ENERGY POLICY FOR EUROPE

IDENTIFYING THE EUROPEAN ADDED-VALUE

CEPS TASK FORCE REPORT

CHAIRMAN: KNUD PEDERSEN
Vice President, DONG S/A and Former Deputy Director-General, Danish Energy Authority

RAPPORTEURS: ARNO BEHRENS
Research Fellow, CEPS

CHRISTIAN EGENHOFER
Senior Research Fellow, CEPS

CENTRE FOR EUROPEAN POLICY STUDIES

BRUSSELS

This report is based on discussions in the CEPS Task Force on "Energy Policy for Europe: Identifying the European Added-Value". The Task Force met several times between October 2006 and September 2007. Participants in this CEPS Task Force included senior executives from a broad range of industry – including energy production and supply companies, energy-intensive industries and service companies – and representatives from business associations and non-governmental environmental organisations. A full list of members and invited guests and speakers appears in Annex 2.

The members of the Task Force engaged in extensive debates in the course of several meetings and submitted comments on earlier drafts of this report. It reflects the general tone and direction of the discussion, but its recommendations do not necessarily reflect a full common position agreed among all members of the Task Force, nor do they necessarily represent the views of the institutions to which the members belong.

ISBN 13: 978-92-9079-773-9

Centre for European Policy Studies
Place du Congrès 1, B-1000 Brussels
Tel: (32.2) 229.39.11 Fax: (32.2) 219.41.51
E-mail: info@ceps.eu
Website: http://www.ceps.eu

CONTENTS

PREFACE

After decades of hesitation, EU member states have started to realise the possible benefits or even the need for a more integrated EU energy policy. New challenges, notably climate change, but also the resurfacing of old ones such as security of supply or the need to face up to increasing global competition, have triggered this new thinking. The first-generation EU energy policy remains schematic and will continue to be so for the time being. Now that agreement has been reached on the principle, additional steps will be needed to identify what can and should be done at EU level and what is best dealt with by member states.

Against this background, CEPS has launched a Task Force that has brought together a broad range of stakeholders, including senior executives from numerous different industries, e.g. energy production and supply companies, energy-intensive industries and service companies, and representatives from business associations and non-governmental environmental organisations. The members of the Task Force engaged in extensive debates in the course of several meetings. During these meetings, the Task Force members had ample opportunity to enter into discussion and to exchange views with senior officials from the EU institutions and international organisations.

This CEPS Task Force Report has been designed as a general reflection on the added-value of an EU energy policy. While identifying the main direction and principles of the new EU energy policy, the report looks to future priorities and highlights key measures that will be crucial for a successful energy policy for Europe.

I want to thank the members of the Task Force for their active and positive contributions throughout the meetings. Although each member endorses the general content of the report, one should not conclude that all members subscribe to every sentence of the text.

Knud Pedersen
Chairman of the CEPS Task Force
Vice President, DONG Energy S/A

EXECUTIVE SUMMARY

The European Union's debate on energy policy has entered a new and crucial stage. Since the adoption of a new integrated EU climate and energy policy at the European Council of 8 and 9 March 2007, the European Commission has tabled various proposals for its implementation, the most recent on 23 January 2008. These proposals are critical in further shaping the emerging contours of this new energy policy for Europe, which aims to address Europe's three main energy policy challenges: to combat climate change and guarantee security of energy supply while ensuring the competitiveness of European industries.

This CEPS Task Force Report contributes to the emerging contours of this new 'first-generation' EU energy policy. Its purpose is two-fold: i) to identify those policy areas that are expected to benefit most from deeper EU integration, i.e. where the European 'added-value' is expected to be biggest, and ii) to formulate recommendations on how such integration could be achieved in practice.

The report is organised as follows. The Executive Summary with Key Messages and Recommendations is followed by 7 sections. The first two sections set out EU energy policy challenges and identify possible areas where European 'added-value' is the greatest. Section 3 introduces the concept of robustness and lays out the need for energy policy indicators. Section 4 analyses the external dimension of a European energy policy, before sections 5 and 6 examine costs and benefits and strategies for cost-minimisation. Finally, section 7 identifies the 'added-value' on the concrete example of renewable energy sources. A list of members of the Task Force can be found in Annex 2.

I. Key Messages

Elements of a 'first-generation' EU energy policy

1. On balance, the new integrated EU climate and energy policy as proposed by the European Commission and endorsed by the European Council is a visionary and yet realistic approach for dealing with EU energy policy challenges. It points in the right direction to what is to become a 'first-generation' EU energy policy. However, a principal challenge is to identify the EU's 'added-value' and to specify how to adapt EU tools, policies and institutions accordingly, in the absence of comprehensive EU energy policy competencies. The agreement on the Lisbon Treaty, including new EU competencies on energy policy, adds a new dynamic as this establishes for the first time an energy chapter and streamlines decision-making.
2. This report has identified four priority areas where the EU's added-value has the greatest potential:
    - Complete the internal energy market with adequate policy harmonisation,
    - Develop a 'European concept for security of supply', including better policy coherence,
    - Build stronger external energy policy capability and
    - Push the development and deployment of energy technologies.

    These four priorities can be reinforced by EU collection and interpretation of information and data, and complemented by institutional adaptation. Finally, the report puts the EU emissions trading scheme (ETS) and renewables policies to the test against the new integrated EU climate and energy policy.
3. The report finds also that a critical element for this added-value to materialise will be policy coherence, especially between the internal energy market and national policies. There is currently no mechanism to ensure a systematic assessment of impacts of national energy policies and measures on the EU. More coherence between EU and member state policies could be expected if the European Commission develops a number of (energy policy) indicators against which to benchmark member states' policies in the context of the Strategic EU Energy Review. In order to be effective, such an assessment would almost certainly need to be made *ex-ante*, which would mean some sort of advance consultation by member states, if not notification to the European Commission.

Reaping the benefits of the internal energy market

4. An effective internal market for energy is a necessary precondition for EU energy policy objectives. It provides a stable and predictable regulatory framework designed for equal opportunities and equal treatment of all market participants. It increases efficiency of the sector while bringing down costs of goods and services by better resource allocation while internationally increasing the economic and political weight of the EU and its member states. At the same time, it will minimise the costs of the EU's ambitious policy targets. A well-functioning internal market is also a strong base on which to develop robust trade relations, thereby turning dependency into interdependency. This includes, among others, effective unbundling, cooperation of independent national regulators and competition policy and the development of cross-border trade and regional markets. The European Commission's proposal of 19 September 2007 to strengthen existing internal market provisions is an important step towards a better functioning internal market and improved competitiveness. Better infrastructure, more competition and new entrants will increase the efficiency of the market. Long-term contracts can help to improve infrastructure and competition, provided the different parts of the energy value chain, production, transportation, wholesale markets, distribution and supply are efficient and free of distortions. It is important that the revised ETS and the renewable energy sources (RES) Directive are in line with the 3rd liberalisation package, enabling Europe to reap the benefits of the internal energy market.

A European concept for security of supply

5. Member state policies to secure energy needs are increasingly influenced by EU policies, such as on energy market liberalisation, renewables or emissions trading. At the same time, national responses to security of supply may partly be incompatible with the security of supply interests of other member states or the EU as a whole. Such inconsistencies can be overcome by a 'European concept for security of supply'. This concept could include tools to ensure policy coherence (e.g. energy policy indicators), foster network development, assess investment needs and deal with political risks and with supply disruptions (e.g. solidarity obligations). A European concept of security of supply would also need to include EU-wide policy measures to address typical market failures, i.e. i) long-term

technology development, demonstration and deployment of breakthrough energy technologies; ii) energy efficiency in domestic sector; iii) consumer behaviour; and iv) possible risks to energy security. Such a concept would need to take into account that the EU is largely dependent on imports from regions where market forces do not work and economic decisions on whether to explore, produce or sell energy are largely linked to political considerations.

External energy policy, oil and gas

6. In terms of energy security, the EU could realise added-value by developing partnerships with supplier countries, acknowledging EU legitimate interests in reciprocity and security of supply and suppliers' legitimate interest in security of demand and stable investment conditions. Such partnerships can establish a real interdependence between the EU and its suppliers, and thereby can enhance predictability and foster investment.
7. There is a need to better integrate energy policy and foreign policy. This is best done by institutionalising dialogues with producer countries, by using existing tools available, such as the European Neighbourhood Policy or trade or development policies. By using all available instruments and fora, the EU can effectively support companies in gaining access to reserves.
8. There is no doubt that an EU unified about energy issues would gain negotiating power on the world stage. A first move towards the EU 'speaking with one voice' could consist of mandates given by member states to the European Commission to negotiate specific energy-related issues within particular EU policies. Such a mandate, however, would require the European Commission, notably the different Directorates-General that are involved in energy, competition and external policy, to integrate their different approaches vis-à-vis third countries into a unified negotiation position. Progress could also be made in the Council of the European Union by improving the coordination between the relevant Working Groups dealing with EU security policy and the Working Group on Energy, or by appointing a Special Representative for Energy in the Council Secretariat.
9. Oil and gas will remain a central part of a coherent energy policy for Europe. Although EU reserves for oil and gas dwindle, there is still significant indigenous oil and gas production in the EU and the EEA

(European Economic Area), notably in Norway. Constant improvements in exploration technologies and extraction methods alone have helped to dramatically increase recovery rates. The EU added-value is to improve conditions for investment and trade both within the EU and externally, including infrastructure development. Ensuring that investments are made and that the considerable remaining potential is fully exploited requires an enduringly competitive regulatory regime. Stimulation of investments both in Europe and globally will contribute to meeting the requirements of a secure, sustainable and competitive energy market.

10. External energy policy cannot secure Europe's energy needs alone. It should be complemented by establishing a fully integrated and competitive internal market and by diversification of the energy mix with more energy from renewable sources.

Research, development and demonstration of energy technology (RD&D)

11. The EU needs to step up its effort in the global race to develop more climate-friendly technologies. The EU's leading role in combating climate change and in increasing energy efficiency offers opportunities for a worldwide advance in related technologies and patents, and for new domains of excellence and export worldwide. The European Strategic Energy Technology Plan (SET-Plan) is a step in the right direction aimed at boosting a clean technology sector currently characterised by high costs, market barriers and underinvestment.

12. Diffusion of existing and cleaner and more efficient technologies can have a dramatic effect on reducing both GHG emissions and import dependence. Such technology diffusion primarily depends on incentives for investment and the related regulatory environment. Government R&D support is required, especially in areas where the economic return to industry is uncertain due to the level of economic risk, very long lead times to mass market or both (e.g. some renewables, fusion, hydrogen, battery technology, CO_2 sequestration). In these cases, government support is needed for both basic research and demonstration, i.e. the pre-commercial development of new technologies.

13. To create added-value, the EU could define technologies that should receive (temporary) support for R&D and demonstration (RD&D). Examples are the stimulation of early market investments in the

development of low-emissions technologies for fossil-fired power plants (CCS) or second-generation biofuels for the European transport sector. In addition, a principal EU role is to ensure better coordination of industry and member states' RD&D programmes and dissemination of results. The SET-Plan aims at facilitating these coordination efforts. Another effective means to support industry RD&D has been EU support for industry commitments such as industry technology platforms.

Backing up EU priorities through data collection and interpretation of information & institutional adaptation

14. There are economies of scale for collection and interpretation of information and data at the European level as it is already acknowledged through the existence of EU bodies such as the European Energy Agency (EEA), Eurostat or the proposed Office of the Energy Observatory. While ample energy and climate data on energy demand, supply or investment needs are available within the EU, the International Energy Agency (IEA) or other international institutions exist, such data are not necessarily geared towards EU-level policy-making. Key economic data on such issues as future investment or infrastructure, e.g. grids, pipeline, liquefied natural gas (LNG) or storage facilities, will not only facilitate decision-making but also increase transparency.

15. A Europe that can act needs suitable institutional arrangements. In many cases, coordination will suffice, especially if backed up by good information and data. In few cases, however, there will need to be new institutional arrangements or bodies such as the proposed agency for the cooperation of national energy regulators or an energy coordinator in the Council secretariat.

Putting the 'new EU energy policy' to the test: Long-term greenhouse gas emissions and renewables targets

16. Long-term binding targets are a good means to express a vision of what the EU and its member states hope to achieve. They thereby affect long-term investment decisions, provided these targets can be met with available technologies and within the necessary deployment times. Yet, they also pose risks, notably in the form of market segmentation, rent-seeking, loss of credibility in the case of apparent non-achievement or reversal of policies, which could lead to stranded investment. If they are too detailed, targets will deteriorate into

planning tools, overriding market incentives and reinforcing rent-seeking. The EU targets and policies thus require a rigorous assessment as to their potential impacts, notably environmental effectiveness, cost-effectiveness, competitiveness impacts and implications for the internal market.

17. Investment will follow if 'high-level' targets are translated into credible and realistic implementation strategies by member states, flexible enough, in particular, to adapt to the diversity of members states, due to i) their present energy mix, ii) the carbon emission intensity of their energy production, iii) the potential of renewables offered by their geography and iv) the background of existing energy savings and energy efficiency improvements. The 'added-value' of the EU is to ensure that EU targets are formulated and implemented in such a way that they are coherent with other EU policies, especially the internal market, and ideally mutually reinforcing. Such coherence is best ensured if the EU develops a joint framework within which member states can implement policies to reach targets. National targets will always constitute a challenge to the EU as they can distort the internal market and reduce efficiency gains. Tradable permit schemes such as the EU ETS are a possible tool to address this weakness. But as the EU ETS has shown, in order to make such schemes work, a comprehensive EU framework including sanctions is needed.

The EU ETS

18. We can expect that the EU emissions trading scheme (ETS) will remain the most important, yet not the only policy tool to implement the EU's 'high-level' GHG emissions targets. The EU ETS has so far not been able to live up either to its environmental nor to its economic promise. The principal shortcomings have been a lack of environmental delivery, distortions to competition, wealth transfer and unnecessary transaction costs. As a result, there were no clear signals for investment, a critical tool to meet the three EU energy policy objectives on climate change, energy security and competitiveness. The proposal for the review of the ETS has tried to address these shortcomings and has notably concentrated on increasing predictability. The added-value of the EU is to ensure environmental effectiveness, non-discrimination, more harmonisation and the reduction of administrative and transaction costs.

Renewable energy: Utilising the experiences from the ETS

19. Ideally, targets that are either set at EU level or derived from it should be implemented within an EU policy framework to avoid the disconnect between the market and regulation. Given the ambition level of renewables, it will however be indispensable that member states retain some flexibility to achieve renewables targets as a result of resource endowment, structure of the energy sector, economic conditions, and differences between past policies or member state political preferences. Cost-effectiveness can be achieved by allowing flexibility between renewables and efficiency. In principle, member states' flexibility (and by extension cost-effectiveness) are best guaranteed if targets are expressed for renewables as a whole as a percentage of total energy consumption. At the same time, it is important that EU member states with a high natural potential for renewable energy expansion will not also be asked to foot the bill on behalf of the entire EU. It is critical to the long-term longevity of the EU's renewable ambition that burdens are shared in a way that is perceived to be fair by member states. In other words, the *financial* costs of realising the required physical expansion of renewable energy in the EU as a whole must be fairly shared among all EU member states.

20. If member states are accorded too much discretion, there is a risk of market fragmentation or the development of niche products in member states, which would increase costs by the need to adapt fuel-consuming applications. The EU's recent proposal for a new renewable energies Directive will need to be consistent with other energy policy measures. Specific attention should be given to the third legislative package and the EU ETS; consistency with these policies would imply that cross-border trade should be facilitated. The ideal situation would be an EU-wide EU renewables strategy, including one EU support mechanism such as a real possibility for cross-border trade in green certificates. Without the benefit of the internal market, European customers would probably face a much higher financial burden in fulfilling the 20% target. Cross-border trade would also ensure that technology development is truly European and thus lead to a better utilisation of renewable energy sources in Europe, including the possibility of ensuring uptake of both biomass and wind energy resources. To date, however, member states have applied radically different support schemes and there is

no consensus between member states on a possible EU scheme. Similarly, as long as financial support is provided from within a country, policy-makers will find it difficult to accept that national resources are used for investment in another member state. This may however change as markets and economies further integrate, as has been the case in the Nordic power market. As an intermediate step, the EU could develop the necessary steps to enable future integration of support schemes and minimum coherence, such as the preparation of harmonised guarantees of origins and grid integration but also of member states' support schemes.

21. To ensure that the new EU legislation on renewable energy sources, i.e. the RES Directive, is able deliver discernible European added-value, it should explicitly build on the internal energy market. That is, it should provide more than the sum of 27 national schemes. At the same time, the Directive should not undermine efficient national support schemes. One way to achieve this, as proposed in the draft RES Directive, is to require member states to measure their target compliance using guarantees of origin as the 'common currency'. Proper rules to ensure that one member state is not subsidising the target achievement in another are foreseen.

II. Recommendations

1. The European Commission should develop a number of energy policy indicators against which to assess member state energy policies to ensure coherence between EU and member states energy policy measures, by using the bi-annual Strategic Energy Review.
2. Member states must agree on effective solution(s) to 'unbundling'.
3. The European Commission has acknowledged the role of long-term contracts between external producers (i.e. upstream) and companies supplying gas to customers in the EU. However, the potential role of long-term and market-based downstream bilateral supply agreements in encouraging investment (by ensuring more predictable prices) should also be taken into account, as long as they are in compliance with EC competition law.
4. The EU must develop an agreed common 'European concept of security of supply', based on an effective internal market but complemented by tools to ensure policy coherence (e.g. energy policy indicators) and measures to improve (strategic) network

development and investment, deal with political risks, supply disruptions and solidarity measures.

5. The European Commission should continue to develop further EU energy security externally through partnerships with supplier countries, notably acknowledging EU legitimate interests in reciprocity and security of supply and suppliers' legitimate interest in security of demand and stable investment conditions. The European Commission and member states must acknowledge that such partnerships can establish a real interdependence between the EU and its suppliers, and thereby can enhance predictability and foster investment.
6. To facilitate such partnerships by more coherence in the external communication of the EU, the European Commission must take the necessary organisational steps to ensure that the various directorates-general dealing with energy speak with one voice.
7. In addition, member states could give mandates to the European Commission to negotiate on specific energy-related issues within particular EU policies with the aim of gradually building an EU that 'speaks with one voice'.
8. In order to better integrate energy and external policy and coherence in general, the EU must improve capacity (in all institutions) to deal with these cross-cutting issues.
9. The EU and member states should put more emphasis on developing EU/EEA indigenous oil and gas resources by facilitating technical progress with respect to innovative and cost-effective techniques to improve recovery rates.
10. The EU and its member states should pool their R&D efforts in the development of more efficient and new decarbonised energy technologies, especially in areas where there is EU added-value, i.e. because of economies of scale or cross-border externalities.
11. The EU and its member states should promote the speedy implementation of the SET-Plan both in terms of R&D and demonstration (i.e. commercialisation) of promising technologies, such as CCS and certain renewables, given the significant potentials for reducing greenhouse gas emissions and import dependency.
12. In formulating further EU energy policy targets, the EU and its member states should be guided by available technologies and their deployment times.

13. The European Commission should rigorously assess national implementation policies to reach EU legally binding targets as to whether they are likely to achieve the target and also in terms of their effects on the internal market. If needed, the EU should develop EU-wide frameworks to achieve such targets while minimising negative impacts on the internal market. The phasing-in of a European green certificates-market on top of existing national support policies is one option.

14. A binding target at EU level should be applied to all member states, but the principle of differentiation between member states' targets for renewables, energy efficiency and other low-carbon technologies – justified by material differences between member states – should be valid for all low-carbon technologies. For biofuels, care has to be taken to ensure harmonised fuel quality standards throughout the EU instead of creating a fragmented market, which would cause high costs for both infrastructure and car technology. This approach should aim to maintain as broad a technology portfolio as possible and avoid market fragmentation brought about, for example, by different fuel specifications.

1. INTRODUCTION

In an effort to come up with a comprehensive and ambitious response to European energy and climate change challenges, the European Union continues to broaden the reflection about its future energy systems - taking into account globalisation, increasing market liberalisation, environmental pressures, technological challenges, and the growing import dependency from politically unstable regions. Following the publication of its Green Paper "A European Strategy for Sustainable, Competitive and Secure Energy" (European Commission, 2006a) in early March 2006, the European Commission tabled a major energy policy package on 10 January 2007, entitled "Energy Policy for Europe",[1] which was accompanied by a number of sectoral policies to implement the overall strategy.[2] After discussions in the EU Energy and Environment Councils, the EU heads of

[1] Other major contributions to this debate include the climate change Communication and background paper on "Winning the battle against climate change" (European Commission, 2005a; 2005b), the Green Paper on Energy Efficiency (European Commission, 2005c), the Energy Efficiency Action Plan (European Commission, 2006b), the works of the High Level Group on Competitiveness, Energy and the Environment (European Commission, 2006d; 2006e) and the enquiry into the energy sector by DG Competition of the European Commission in 2006.

[2] COM(2007) 1: An Energy Policy for Europe; COM(2007) 2: Climate change Communication "Limiting global climate change to 2° C"; COM(2006) 845: Biofuels progress report; COM(2006) 848: Renewable energy road map; COM(2006) 849: Report on progress in renewable electricity; COM(2006) 841: Prospect for internal gas and electricity market; COM(2006) 846: Priority interconnection plan; COM(2006) 847: Strategic energy technology plan; COM(2006) 843: Sustainable power generation from fossil fuels: Aiming for near-zero emissions from coal after 2020; COM(2006) 844: Nuclear illustrative programme.

state and governments at their European Council meeting of 8 and 9 March 2007 (European Council, 2007), by and large endorsed the European Commission's strategy including:

1) A binding absolute emissions reduction commitment of 30% by 2020 compared to 1990 conditional on a global agreement,[3] and a "firm independent commitment" to achieve at least a 20% reduction by 2020. At the same time, the EU advocated that industrialised countries reduce their emissions collectively by 60% to 80% by 2050 compared to 1990. The European Parliament in its resolution has insisted that the EU should unilaterally commit to 30%.
2) 20% reduction of primary energy consumption by 2020 compared to projections;
3) A binding target of 20% of renewable energy in total energy consumption by 2020;
4) A binding minimum target of 10% biofuels of all transport fuels by 2020;
5) The development of a European Strategic Energy Technology Plan;
6) An endorsement of the European Commission's carbon capture and sequestration policy.

In the follow-up to the spring 2007 Council, the European Commission has tabled various proposals to implement the European Council decisions. A Strategic Energy Technology Plan[4] was published in November 2007, focusing on specific technologies that may help to achieve the 2020 commitments. On 23 January 2008, the Commission presented a whole package of proposals,[5] containing an update of the EU emissions trading system (ETS),[6] binding national targets for the reduction of

---

[3] Provided that other developed countries commit themselves to "comparable" reductions and economically more advanced countries to contributing "adequately" according to responsibility and capabilities.

[4] COM(2007) 723: "Towards a low carbon future – A European Strategic Energy Technology Plan".

[5] See also COM(2008) 30: "20 20 by 2020 – Europe's climate change opportunity".

[6] COM(2008) 16: "Proposals for a Directive of the European Parliament and of the Council amending Directive 2003/87/EC so as to improve and extend the EU greenhouse gas emission allowance trading system".

greenhouse gas (GHG) emissions outside the EU ETS until 2020,[7] binding national targets for increasing the share of renewable energy sources in final energy consumption in 2020,[8] proposals on biofuels including environmental sustainability criteria,[9] new rules to stimulate CCS,[10] as well as new state aid rules. The adoption of the package by member states is foreseen for the end of 2008, at the earliest.

Priorities for international cooperation have been formulated in the Action Plan that is annexed to the European Council Presidency Conclusion of 8-9 March 2007. Priorities in regard to developing countries are: bilateral energy dialogues with China, India, Brazil and other emerging economies, focusing on the reduction of GHG emissions, energy efficiency, renewables and low-emission energy technologies—notably CCS; enhancing energy relationships with Algeria, Egypt and other oil-producing countries in the Mashreq/Maghreb region; a special dialogue with African countries on energy and the use of Community instruments to enhance decentralised renewable energies in particular, and generally energy accessibility and sustainability in this region, as well as energy infrastructure of common interest; and promoting access to energy in the context of the UN Commission on Sustainable Development (UNCSD).

At the same time, the EU continues to play a key role in the development of emerging carbon markets, both in running the world's largest GHG allowance market in the form of the EU ETS and in generating the significant demand for clean development mechanism (CDM) and joint implementation (JI) credits.

---

[7] COM(2008) 17: "Proposals for a Decision of the European Parliament and of the Council on the effort of Member States to reduce their greenhouse gas emissions to meet the Community's greenhouse gas emission reduction commitments up to 2020".

[8] COM(2008) 19: "Proposals for a Directive of the European Parliament and of the Council on the promotion of the use of energy from renewable resources".

[9] Included in the proposal on renewable energy resources.

[10] COM(2008) 13: "Supporting Early Demonstration of Sustainable Power Generation from Fossil Fuels", and COM(2008) 18: "Proposals for a Directive of the European Parliament and of the Council on the geological storage of carbon dioxide and amending Council Directives 85/337/EEC, 96/61/EC, Directives 2000/60/EC, 2001/80/EC, 2004/35/EC, 2006/12/EC and Regulation (EC) No 1013/2006".

To contribute to this debate, the CEPS multi-stakeholder Task Force on Energy Policy for Europe has produced this CEPS Task Force Report,[11] which attempts to develop the key elements for a EU energy policy framework. Although this report focuses mainly on the internal aspects, i.e. what the EU can implement internally, it nevertheless takes into account the external dimension, as this is an important factor in EU energy policy.

This CEPS Task Force Report concentrates on six key areas that are discussed in chapters 3 to 8.

1. The starting point for an EU energy policy framework is to identify the added value of a European energy policy for the member states. A truly integrated EU energy policy framework will need to be more than just pasting together a number of largely isolated sector-specific policies. Putting together a package of unconnected policies risks producing a sub-optimal outcome that neglects the inevitable interaction between policies (chapter 3).
2. The report introduces the concept of 'robustness' with regard to EU energy policy, referring to its ability to deal with risks associated with developments in the energy market. At the heart of this concept is the EU internal energy market. The report makes a number of suggestions to enhance robustness both by improving the functioning of the internal energy market and by complementary government action. It highlights the need for robustness indicators and calls for improvements in the collection and interpretation of information and data, both at the EU and member state level (chapter 4).
3. To make progress towards a situation where the EU can speak with one voice, this report ponders the possibilities for the EU to assess member state external energy policy actions as to their impacts at the EU level and to create more consistency and coherence between EU and member state policies. Key issues for international energy cooperation and diplomacy will be energy efficiency, investment in production and infrastructure, market access and climate change policies (chapter 5).

---

[11] In March 2007, the Task Force published an Interim Report containing key messages and recommendations in the lead-up to the Spring 2007 European Council (available for free downloading on the CEPS online bookshop http://shop.ceps.eu).

4. The achievement of both energy security and climate change targets is likely to entail costs, i.e. it will require paying a premium. In order to allow for long-term competitiveness of EU industry - one of the three overall objectives of the EU - the report discusses how costs can be minimised (chapter 6).
5. This leaves the questions of which is the best instrument to deal with security of supply and long-term sustainability and whether market-based or non-market based instruments are the best options (chapter 7). The same chapter sets out the conditions under which targets can work. It finds that EU targets are useful in expressing a vision, but they need to be backed up by credible and realistic implementation strategies.
6. The concluding chapter 8 outlines a roadmap for the gradual establishment of an EU framework for renewable electricity support.

The main report is preceded by an Executive Summary including Key Messages and Recommendations. We focus in particular on 'robustness indicators' and their importance for the bi-annual Strategic Energy Review; the next steps to ensure effective linkage between competitiveness, energy security and sustainability; and long-term targets within a EU framework for renewable electricity support.

2. CHALLENGES IN EUROPEAN ENERGY AND CLIMATE CHANGE POLICY

The current energy policy discussion attempts to address three principal long-term objectives: i) security of supply, notably growing import dependence, ii) the need to drastically reduce greenhouse gas emissions in the interest of the environment, while iii) maintaining the EU's international competitiveness. The particular challenge is to steer the EU economy towards a more secure and sustainable path while not jeopardising the competitiveness of EU industry and further improving the functioning of EU energy markets.

This has to be achieved against the background of a **more difficult and risky international environment** than in the past.

- European energy demand is increasingly rivalled by demand from emerging economies, especially in Asia. Without new developments on the supply-side, increasing world energy demand will lead to higher prices in the medium and long term posing a real threat to European energy security. The International Energy Agency concludes that "ensuring reliable and affordable supply will be a formidable challenge" (IEA, 2007).
- Oil and gas reserves in the EU are dwindling (see BP, 2007). In some countries, which could fill the gap, EU industries have little or no access to reserves. Access to energy reserves becomes a critical issue for the EU and other consumer countries.
- Energy industries in supplier countries are subject to extensive government interference, and do not function in a competitive market framework. To a large extent, energy production and export companies are state-owned monopolies. This adds to the fears that energy will increasingly be used as a political weapon. Government-regulated

investment policies also raise doubts about the level of future investments and their effects on production and price levels. In the past, many supplier countries have proven unable to increase production, adding to the pressure on market prices (e.g. Riley, 2006). In addition, security of supply is threatened by political instability of exporting regions.

- Many reserves will take years to develop due to problems of access, investments and physical conditions. A prolonged tight market might increase political tensions and foster a form of 'resource nationalism'. On top of this, the absence of visible progress towards a global climate regime increases uncertainty for investors.
- On the positive side, there is a strong and ambitious commitment to energy efficiency and conservation in the EU and most other OECD countries and many technological options – such as new techniques for exploration and extraction, renewables, nuclear or clean coal – exist that could over time address the collective challenges of security of supply, climate change and economic competitiveness.
- The internal market for electricity and gas, which currently consists of different national or regional markets, is still developing and is far from functioning smoothly. Without an efficient internal market, the EU/EEA response is likely to lack efficiency, which is normally imposed by market discipline, and competitiveness, and member states will feel compelled to resort to national solutions, further undermining the internal market. The lack of transparency in the price formation mechanisms – which is also a crucial condition and an indicator for the functioning of the internal market – inhibits new entry, at least in some member states.

In addition, the following sector-specific challenges can be identified:

- The supply of **oil** in the EEA and OECD countries has largely been secured by existing infrastructure, reinforced by security measures such as the IEA strategic stocks and demand restraint measures. However, Europe and all other major consumer regions are faced with declining domestic oil extraction rates, leading to increasing import dependency and stronger competition between oil importers. At current production rates, the EU-25 will have exploited all its proved reserves by the end of 2014 (BP, 2007). On the other hand, global estimates of how much recoverable oil remains have consistently increased over time. BP (2007) estimated global oil reserves to amount to 1.2 trillion barrels at the end of 2006. At the current rate of global production, these remaining reserves would last for another 40.5 years. This reserves-to-production (R/P) ratio has remained rather constant when compared with 1996 and 1986 levels of 41 and 39.8 years, respectively. However, appropriate investment is the key if future production is to succeed in matching demand. Additionally, undiscovered conventional resources are estimated to amount to 880 billion barrels (IEA, 2006a). In the long-run, non-conventional oil resources (e.g. oil sands, gas-to-liquids, coal-to-liquids, oil-shale, etc.) will play an increasing role in global oil supplies. With estimated reserves of at least 1 trillion barrels, they are expected to contribute about 8.5% to global oil supplies by 2030 (IEA, 2007). Moreover, these resources are mostly located outside the current oil-producing countries. Whether non-conventional alternatives will be commercially available will depend on world oil prices and technology/infrastructure development.

- As for natural **gas**, whatever the expected increase of the LNG market, the EEA will increasingly depend on gas pipe supplies coming from very few countries. Whereas over 80% of the global natural gas reserves of 181.5 trillion m^3 (BP, 2007) are located close enough to Europe to allow for pipeline transport (Müller, 2007), Europe lacks the infrastructure to tap resources in the Middle East, which has the largest proven reserves (over 40% of global reserves). Over 80% of Europe's natural gas imports come from three countries (European Commission, 2006b), where the gas market is tightly controlled by governments. In this respect, fears of 'gas cartels' or of energy being used as a political weapon do not seem unfounded. Similarly, there is a risk of a lack of investment in exploration, production and transportation, despite reserves being abundantly available in areas surrounding Europe. If gas is unable to take a larger share in power generation, it will not be able to live up to the expectation that it can serve as a 'bridge' to a low-carbon economy and may even become a 'sunset' industry.[12] The future carbon price will also have an impact on future gas markets.[13]

---

[12] Unless a combination of demand reduction, new and renewable energies and nuclear power can fill the gap, the substitute would be coal, resulting in higher GHG emissions (J. Stern, 2006).

[13] Most gas companies have adjusted their growth expectations downwards after the first carbon market experience, due to competition of gas with coal and nuclear for power generation.

- Issues for the **electricity** sector are continuity and reliability of supply. Risks include electricity blackouts due to ageing infrastructure and a lack of investment in networks, especially cross-border capacity, erosion of reserve capacity, public opposition to new investment projects in generation and transportation, bad regulation and increasing the market share of incumbent companies in domestic generation. A particular aspect is the slow progress in liberalisation and integration of the electricity market. Markets in 'transition' create their own risks such as erosion of reserve capacity. Such issues are likely to disappear, once the market functions properly.
- On top of this, the EU faces the long-term **sustainability** challenge, namely, the problem of climate change. It is generally assumed that in the long term industrial countries will need to reduce emissions by up to 60% or more by 2050, not to mention further reductions beyond by 2050.[14] Given that within the EU, 80% of all emissions are related to fossil fuel burning in the energy, transport, household and industrial sectors, energy policy will increasingly be constrained by climate change objectives. While near zero carbon energy or possibly fusion will ultimately be essential to meet the climate change challenge, the present focus is on how to reduce GHG emissions from fossil fuels such as coal, oil and gas, which continue to dominate the EU's energy mix. The principal obstacle facing the EU is the absence of a comprehensive global climate-change agreement that would provide the necessary certainty for investors when making investment decisions. In its absence, it is up to governments both at member state and EU level to provide signals to induce investors to make more carbon-friendly investments, bringing Europe onto a trajectory towards a near-zero carbon economy while neither jeopardising security of supply nor competitiveness.

---

[14] For a stabilisation at 450ppm CO_2e, the Stern Review (N. Stern, 2006) assumes a global peak of GHG emissions already by 2010 with a 6-10% decline thereafter. For stabilisation at 550ppm CO_2e and if emissions peak in 2020, the estimate annual declines are around 1-2.5% annually afterwards.

3. THE EUROPEAN ADDED-VALUE

A European energy policy should be placed in the overall context of the EU, which is based on an internal market and the principles of a competitive market economy, free from distortions of restrictive agreements, the abuse of dominant positions or state aid. At the same time, the EU's treaties list other important objectives, including notably sustainability, which are sometimes mutually reinforcing while at other times necessitating trade-offs. EU energy policy equally needs to be seen in the context that - up to now - member states have granted only very limited energy *policy* competencies to the EU while at the same time ceding significant powers in other policy areas such as the internal market, competition policy, the environment or research. A principal objective of this emerging EU energy policy framework therefore is to identify the *European added-value* to national energy policy-making, not to mention the legal basis. The principle of subsidiarity dictates that an EU role is warranted where EU action genuinely has benefits. Otherwise, action should be retained at member state or international level (e.g. the IEA, UN or G8).

3.1 The internal energy market

The further development of the EU's internal energy market for gas and electricity is often mentioned as one of the principal advantages of a European energy policy. It provides a stable and predictable regulatory framework designed to ensure equal opportunities and equal treatment of all market participants. In this context, on 19 September 2007, the European Commission made another attempt at completing the process towards an internal energy market in the EU by proposing a number of measures to complement the existing rules. Proposals contained in the 'third package of legislative proposals' include the separation (unbundling) of production and supply from transmission networks (this would also apply to companies from third countries), the establishment of an agency for the cooperation of National Energy Regulators, improved market transparency and increased solidarity between member states. While a pan-European energy market is under development – reinforced by new and additional competencies included in the Treaty of Lisbon – member states will likely continue to implement national policies suited for their specific market situation. It is vitally important that member state measures do not delay or hinder cross-border markets from emerging, in order not to undermine the potential efficiency gains from a functioning pan-European market, and avoiding a further re-nationalisation of energy policy.

Box 1. Provisions of the Treaty of Lisbon

Energy issues are referred to on several occasions in the Treaty of Lisbon signed on 13 December 2007 by the EU Heads of State or Government. According to the final text still to be ratified by member states, energy policy will be a shared competence between the Union and its member states in the amended Treaty on European Union (Art. 2 C). In comparison to the previous proposal for a Draft Constitution, it includes additional reference to new challenges, such as climate change and energy solidarity between the member states in case of difficulties in supply (Art. 100). The latter takes into account concerns by certain member states about high energy dependence on one supplier country and the effects of possible disruptions of supply. A specific chapter devoted to energy (Title XX, Art. 176 A) highlights the importance of a functioning internal energy market in line with the need to preserve and improve the environment. Security of supply is the central focus, alongside the promotion of energy efficiency and energy saving and the development of new and renewable forms of energy. Based on this article, the Union shall aim at promoting the interconnection of energy networks. However, any measures to that effect "shall not affect a Member State's right to determine the conditions for exploiting its energy resources, its choice between different energy sources and the general structure of its energy supply", thus leaving member states' flexibility largely untouched.

Source: Treaty of Lisbon amending the Treaty on European Union and the Treaty establishing the European Community, signed at Lisbon, 13 December 2007.

3.2 Beyond the market

In addition to a fully integrated and competitive market, government action will be needed especially for long-term policy objectives. Examples include R&D or the development of new and breakthrough technologies to cope with climate change. Moreover, the EU is dependent on imports from areas where market rules do not apply and economic decisions on whether to explore, produce or sell energy is largely linked to political considerations.

This Task Force has identified a number of areas where European added-value is given in terms of government intervention:

1. On the **demand side**, the promotion of ambitious energy saving and energy efficiency policies could reduce dependence on politically unstable or unreliable countries. Another area is investment in energy efficiency programmes by utilities with a focus on networks upgrade and smart metering systems to give the customers awareness of their consumption through a real time measure. A particular objective should be to limit the use of oil to essential areas such as transport and petrochemical, or a reflection on how to make best use of natural gas.
2. On **the supply side**, support should be given to near-zero carbon technologies such as renewables and carbon capture and storage. This would include especially temporary support mechanisms for renewables electricity or second-generation biofuels[15] to help bring down costs, creating the regulatory framework for carbon capture and storage as well as facilitating pilot and demonstration projects in the area. Ultimately, it should be **up to the markets to choose the appropriate technologies** (be they renewables, carbon capture and storage or nuclear) based on the political objectives of the EU and/or its member states. Choices will most likely be different across member states depending on political preferences, political acceptability and resource endowment.
3. On **R&D, efforts should focus on all energy** technologies, including both demand and supply, with the objective of maintaining or increasing diversification and flexibility of EU and global energy markets but also to reduce GHG emissions from fossil fuels.

[15] Second-generation biofuels here mean biofuels made from wastes, residues, non-food cellulosic material, and ligno-cellulosic material like ligno-cellulose based bioethanols, Fischer-Tropsch diesel and bio-dimethylether. Although conventional biofuels such as pure vegetable oil, biodiesel and ethanol are cheaper, as CO_2 reduction of second-generation biofuels are about the double of those of conventional ones, second-generation fuels have a far better cost-benefit ratio (see Jansen & Bakker, 2006).

Attention will be needed to ensure the **deployment**[16] of technologies to bring down costs – once these have come close to being competitive – while minimising the environmental impact.

Box 2. A European Strategic Energy Technology Plan (SET-Plan)

The SET-Plan was tabled by the European Commission on 22 November 2007. It aims at boosting a clean technology sector, currently characterised by high costs, market barriers and underinvestment. Its purpose is to accelerate the availability of energy technologies and at the same time to engage European industry in the process to help it gain world leadership in this sector. The aim of the SET-Plan is thus to turn technology opportunities into business realities by delivering a new joint strategic planning, a more effective implementation, an increase in resources and a new and reinforced approach to international cooperation. It focuses on key technologies to meet the 2020 targets but ventures beyond available technologies to achieve the 2050 vision towards complete decarbonisation. However, to maintain flexibility in technology development, targets contained in the plan are not binding. Also, the plan does not mention how the targets can be reached and how much it will cost to achieve them.

Source: COM(2007) 723: A European strategic energy technology plan (SET-Plan), 'Towards a low carbon future'.

4. For **oil**, the objective should be to maintain or increase market flexibility both in the EU and globally (e.g. by increasing spare oil production capacity), liquidity (e.g. by preventing oil resources to be excluded from the global market) and diversity (e.g. develop unconventional oil), essentially to reduce transaction costs by improving the functioning of markets.
5. For **natural gas**, the objective should be to improve the functioning of the internal gas market, notably by increasing liquidity both for piped gas and LNG. Liquidity of the market presupposes that it remains attractive for producers to deliver sufficient volumes to the EU and the right incentives for infrastructure investment are in place. In addition, existing bodies such as the Gas Coordination Group could be used to identify possible measures to cope with possible supply

[16] See the presentation of A. Stouge at the CEPS Task Force meeting on 8 November 2006 (Stouge, 2006).

disruptions, including better coordination or harmonisation of national regulations on gas supply and on gas stocks.

6. For the **electricity** sector, continuity and reliability will be enhanced by a more harmonised or even unified management of the European grid, more investment in generation and grids and improved cooperation between transmission system operators (TSOs). These measures, in particular, are a key to foster the regional markets and ultimately European energy markets integration.

The above measures constitute the existing 'EU consensus' of no-regret options to address EU energy policy objectives. Too often, however, such no-regret options fail due to policy inertia, expediency or simply lack of interest. To avoid such failure in the future, the European Commission in its role as Guardian of the Treaties could be given special responsibility for tracking member states and EU progress towards the implementation of these measures.

Röller et al. (2006) show that member states' energy policies remain largely determined by exogenous factors such as availability of domestic resources and geography. Member states have very different starting points, facing different energy challenges. Against such a background of heterogeneity, identifying the 'European added-value' beyond the above consensus will be difficult. On the other hand, if an energy policy for Europe attempts to go beyond the 'lowest common denominator', i.e. the policies and measures that all member states agree with, such a policy will need to submit member states to the test on whether domestic energy policies meet "agreed EU policy objectives".

There are additional possible measures at EU or member state level available to further increase the robustness of the EU energy sector and have a generally positive cost-benefit ratio from an overall social perspective, including cost-efficiency, climate change, other environmental impacts and security of supply. They include public financial support for electricity interconnectors or gas pipelines,[17] common approaches to LNG or gas storage taking into account security of supply, investment in

[17] The provisions of the Lisbon Treaty add an extra aim "to promote the interconnection of energy networks" as well as stating that energy policy should be carried out in a "spirit of solidarity" among member states.

additional supply including renewables, biofuels or nuclear and EU-wide crisis or solidarity measures in case of supply disruption.[18]

3.3 Policy integration

Solving the triple challenge of securing energy supplies, sustainability and competitiveness requires more than just pasting together a number of largely isolated sector-specific policies. Lumping together disconnected policies risks producing a sub-optimal outcome that overlooks the interaction between policies. The recent energy and climate change package of the European Commission, tabled on 23 January 2008, marks a first step to develop an integrated approach towards an EU energy policy.

To date, there are various instruments available to ensure coherence. At the level of the European Commission, they include notably inter-service consultations and integrated impact assessments (IIAs) for all major policy initiatives. Also at Council level, there is a drive towards more policy coherence. As to the effectiveness of these tools, much depends on how they are applied. Efforts are made in both the European Commission and by the Council Secretariat, but initiatives for better coherence have not been without difficulties (e.g. see Renda, 2006; Egenhofer et al., 2006).

There have been examples for integrated approaches. These include the Auto-Oil programme, the European Climate Change Programme and the recent European Commission proposal to tackle CO_2 emissions from cars. Each of these initiatives has constituted considerable effort in terms of scientific input and stakeholder contribution. The Strategic EU Energy Review should draw lessons as to the importance of data, analysis and stakeholder involvement.

Equally important is coherence between EU and member state actions. Given the limited EU competencies on energy policy, member

---

[18] For example, the CPB (2004) has identified the following measures with a positive cost-benefit ratio: extending the size of strategic oil stocks, subsidising biomass in the transport and chemicals sectors, conserving the Groningen gas field, encouraging investments in wind turbines, coal-fired plants or nuclear power replacing gas-fired plants, better incentives for power generation reserve capacity and changes in the regulator of electricity networks. The recent CEPS/ECN study (Egenhofer et al., 2006) has identified, among others, CHP, IGCC, second-generation biomass and technology support as measures with a positive cost-benefit ratio.

states enjoy considerable discretion in this area and it is far from evident that national energy policy initiatives, notably in the area of security of supply are beneficial to the EU as a whole. Examples include the planned 'Nord Stream' and 'South Stream' gas pipelines, some recent equity investment of Gazprom in member states, the lack of a credible renewables strategy or the phasing out of nuclear power without a credible policy to substitute for it. This raises the question of whether future Strategic EU Energy Reviews should include a systematic assessment of EU impacts of national energy policies and measures.

4. The Concept of Robustness of EU Energy Policy

If we follow the European Commission's and the IEA's analysis that energy supplies may become more risky than in the past, the EU would be well advised to put in place a robust domestic system to 'insure' against supply and other energy policy risks. The idea therefore would be to develop an EU policy framework that is both **effective** in achieving EU long-term objectives and sufficiently **robust** to deal with risks. Such a framework would best stress predictability, flexibility, coherence and cost-efficiency, in addition to – if warranted – built-in solidarity or crisis mechanisms. The CEPS Task Force attempted to develop a number of EU energy policy indicators to express robustness of member state and EU energy policies.

Röller et al. (2007) have developed a relatively simple so-called 'energy policy index' (EPI) to assess the performance of member states against the EU's energy policy objectives of competitiveness, security of supply and sustainability (see Table 1). The initial finding is that heterogeneity prevails, mainly due to different exogenous factors, such as geographical location, availability of domestic resources and political preferences. The study finds five different groups of countries, suggesting that there are no clear correlations between scores of the three different energy policy objectives. Hence, member states start at very different starting points, facing different energy challenges.

Table 1. Energy policy index (EPI)

Country		Competitiveness	Security of supply	Environment sustainability
Austria	AT	2.7	3.0	3.7
Belgium	BE	2.0	1.2	1.8
Cyprus	CY	0.5	0.0	2.0
Czech Republic	CZ	2.8	3.1	2.8
Germany	DE	1.9	2.5	3.0
Denmark	DK	3.6	4.0	3.4
Estonia	EE	1.2	3.2	3.3
Spain	ES	1.9	1.9	2.4
Finland	FI	1.5	2.0	4.8
France	FR	0.8	2.6	3.8
Greece	GR	0.8	2.5	2.8
Hungary	HU	2.9	2.1	3.2
Ireland	IE	1.1	0.8	2.4
Italy	IT	2.4	1.9	2.7
Lithuania	LT	3.5	2.4	4.1
Luxembourg	LU	3.9	3.0	2.5
Latvia	LV	2.6	1.5	5.3
Malta	MT	0.0	0.0	1.3
Netherlands	NL	2.6	2.7	3.2
Poland	PL	1.8	4.6	2.6
Portugal	PT	2.3	1.5	3.3
Sweden	SE	2.3	2.7	5.0
Slovenia	SI	4.1	1.9	3.9
Slovakia	SK	2.5	1.7	3.0
United Kingdom	UK	2.9	3.5	3.1

Note: The higher the value of a specific indicator (0 to 6), the better the performance in terms of the criteria defined in the EPI.

Source: Röller et al. (2007).

4.1 Principal EU energy policy risks

There is a long literature on security of supply risks. Based on the International Energy Agency (IEA, 1995), the literature traditionally distinguishes between short-term and long-term risks (see Stern, 2002 and

Luciani, 2004). Short-term risks are generally associated with supply shortages because of accidents, terrorist attacks, extreme weather conditions or technical failure of the grid. This is sometimes described as 'operational security' or 'systems security'. Long-term security is associated with the long-term adequacy of supply, the infrastructure for delivering this supply to markets and a framework to create strategic security against major risks (such as non-delivery for political, economic, *force majeure* or other reasons). The European Commission's 2000 Green Paper on security of supply has identified four risk categories: technical, economic, political and environmental risks.[19]

By thus expanding the narrow concept of security of supply, we can develop a number of EU energy policy indicators to be used as a basis for assessing robustness of member state and EU energy policies. The starting point is the identification of relevant risks. We have identified six:

1. *Import dependence* on producer and transit countries;
2. *Investment risk,* including investments within the EU internal market (e.g. in infrastructure or reserve capacity) and beyond (e.g. in non-EU transportation infrastructure or upstream investment in supplier countries);
3. *Environmental risks* from climate change, regional/local pollution or contamination due to accidents;
4. *Regulatory and political risks* due to inefficient or failed regulation or local market disruptions due to pressure group actions (e.g. opposition against new investment, fuel price protests, etc.);
5. *Risks* associated mainly with *market failure* (e.g. excessive concentration of market power or failure of financial markets); and
6. *Excessive energy prices,* which can originate from any of the above risks or a combination thereof.

---

[19] *Technical risks* include systems failure due to weather, lack of capital investment or generally bad conditions of the energy system. *Economic risks* cover mainly imbalances between demand and supply due to a lack of investment or insufficient contracting. *Political risks* outline potential government policies to suspend deliveries due to deliberate policies or war or civil strife or as a result of failed regulation, which is referred to as regulatory risk. *Environmental risks* describe the potential damage from accidents (oil spills, nuclear accidents) or pollution, including pollution whose effects are less tangible or predictable (e.g. greenhouse gas emissions).

Risks can be further distinguished and grouped in sub-categories as shown in Annex 1. For the purposes of this report, a general categorisation will suffice (e.g. Egenhofer, 2007a).

4.2 Robustness indicators

The assumption is that in order to cope with the risks associated with EU energy policy objectives – competitiveness, security of supply and climate change – both member state and EU energy policies need to match a number of 'robustness indicators'. The CEPS Task Force has identified the following 10 robustness indicators:

1. Share of biggest irreplaceable single *import* source (taking substitution possibilities into account);
2. Share of biggest irreplaceable single *energy* source (taking substitution possibilities into account);
3. Progress towards different EU and national policy targets (e.g. GHG emissions reductions, renewables, energy efficiency and conservation targets);
4. Energy intensity in absolute terms (corrected by climate and other factors) and improvement over time for the economy/domestic sectors;
5. Reserve and excess capacity in generation, interconnections, transportation of natural gas, gasification terminals, gas storage, etc.;
6. Internal market indicators (e.g. relative competition for final consumers, choice of transportation or competition between fuels);
7. A degree of protection of vulnerable consumers against supply disruptions;
8. A set (i.e. reasonable) degree of solidarity measures (excluding for example structural imports);
9. Public and private energy R&D spending; and
10. Tolerable impact of EU-induced energy price increases for EU industry subject to global competition.

While the above-mentioned 'robustness indicators' will need considerable methodological refinement as well as suitable data to become useful, they could become a valuable tool for the European Commission to track performance of individual member states and the EU as a whole or to evaluate the degree to which energy policy objectives are integrated. Such an assessment of member state and EU progress is politically very sensitive, as it will affect member state autonomy. However, a meaningful EU energy policy beyond the status quo - especially in the light of the Lisbon Treaty - will require member states to accept some sort of benchmarking.

4.3 Data requirements

The development of European energy indicators requires considerable improvement in the collection and interpretation of data and information, both at EU and member state level. There are economies of scale for collection and interpretation of information and data at the European level, as testified to by the existence of EU bodies such as the EEA, Eurostat or the proposed Office of the Energy Observatory. While ample energy and climate data on energy demand, supply or investment needs are available within the EU, the International Energy Agency or other international institutions, such data are not necessarily geared towards EU-level policy-making. The collection of key economic data, e.g. on future investment or infrastructure such as grids, pipelines and LNG or storage facilities, will not only facilitate decision-making but also increase transparency.

5. Speaking with One Voice: The Role of the EU's External Policy

One of the recurring themes of the EU integrated climate and energy package is the alleged need for Europe to speak with one voice. Implicitly, this includes coherence between internal and external policies. As early as March 2006, the Green Paper (European Commission, 2006a) identified a "coherent external energy policy" as one of the six EU energy policy pillars.[20] The Green Paper has taken a procedural approach and is putting its faith in the Strategic EU Energy Review to serve as the basis for establishing a common EU vision, which will gradually become a common external voice. In addition, the European Commission had listed an ambitious catalogue of themes for this "common external policy" including amongst others, a *clear policy on securing and diversifying energy supplies* and *reacting effectively to external crisis situations*, in addition to more predictable suggestions such as entering into energy partnerships with producers, transit countries and other international actors, integrating energy into other policies with an external dimension, and using energy to promote development. An important step has been the June 2006 European Council decision to adopt the proposed legal framework for the external energy policy on the basis of the joint paper by the European Commission and the High Representative. Amongst others, this decision foresees the creation of a network of energy correspondents (consisting of representatives by member states and the General Secretariats of both the Commission and the Council) to set up an early-warning system and to

[20] The energy challenges facing Europe need a coherent external policy to enable Europe to play a more effective international role in tackling common problems with energy partners worldwide (European Commission, 2006a).

improve the reaction in case of a crisis. The 2007 Spring European Council finally agreed on an action plan for international energy policy, [21] essentially aiming at better coordination and coherence. A further politicisation of energy issues has also been suggested in a recent report adopted by the European Parliament (2007), which proposes the creation of a post for a High Official of Foreign Energy Policy. Other measures to increase security of supply include the Energy Charter Treaty to be the cornerstone of a common European foreign policy on energy, the creation of a solidarity mechanism to deal with disruptions of supply or infrastructure damage, as well as diversification and increased energy efficiency.

5.1 Preconditions

Although the concept of 'speaking with one voice' expresses the potential *added-value* of the EU presenting a harmonised external position, it faces the dilemma that EU member states pursue different national policies and interests stemming from their heterogeneity and the different starting

[21] The focus is on speeding-up the development of a common approach to external energy policy, involving consumer-to-producer as well as consumer-to-consumer and consumer-to-transit countries, dialogues and partnerships, involving organisations such as OPEC. More concretely, the European Council has formulated the following objectives: i) negotiating and finalising a post-partnership and cooperation agreement with Russia in particular relating to energy issues; ii) intensifying the EU relationship with Central Asia, the Caspian and the Black Sea regions, with a view to further diversifying sources and routes; iii) strengthening partnership and cooperation, building on the bilateral energy dialogues with the US as well as with China, India, Brazil and other emerging economies, focusing on the reduction of GHG emissions, energy efficiency, renewables and low-emissions energy technologies, notably carbon capture and storage (CCS); iv) ensuring the implementation of the Energy Community Treaty, with a view to its further development and possible extension to Norway, Turkey, Ukraine and Moldova; v) making full use of the instruments available under the European Neighbourhood Policy; vi) enhancing energy relationships with Algeria, Egypt and other producing countries in the Mashreq/Maghreb region; vii) building a special dialogue with African countries on energy and using Community instruments to enhance in particular decentralised renewable energies and generally energy accessibility and sustainability in this region, as well as energy infrastructure of common interest; and xiii) promoting access to energy in the context of the UN Commission on Sustainable Development (UNCSD).

points discussed above, such as the degree of energy market liberalisation, differences in the energy mix or levels of diversification, geographical location or even differences in foreign policy objectives.[22]

As a result, the *added-value* for the EU external energy policy has yet to be convincingly defined based on economic and political realities. Nevertheless, there seems to be a consensus on two broad strategies: i) the widening of EU energy markets combined with the reinforcement of energy partnerships with a view to improving the functioning of world markets in energy, and ii) the diversification of energy supplies by source, geographical origin and transit route.[23] The principal EU role therefore will be to improve the basic conditions under which companies and member state policies operate.[24] The European Commission's announcement of its intention to include "major energy chapters in its relations with neighbouring countries" as well as to put energy issues on the agenda of every summit with third countries is an important step towards an increasingly common external energy policy. From an EU perspective, key issues will be energy efficiency (as this reduces global demand and GHG emissions), investment in production and infrastructure (to increase competition and ensure adequate supply), market access (for European and international energy companies) and climate change policies (to reduce global greenhouse gas emissions) as well as other environmental and safety issues, especially in regard to nuclear energy.

---

[22] For example, there are very different views across the EU on how to deal with Russia.

[23] See Paper by the European Commission and the Council Secretary General/High Representative in the summer 2006, and the European Commission's contribution to the Lahti European Council (European Commission, 2006f).

[24] Promotion of principles of the internal energy market in bilateral and multi-lateral fora, combined with improved interconnections are likely to work as long as the EU's partners see benefits from this. This can be expected to work within the context of relations within the EEA, the EU neighbourhood countries, other associated countries and within international organisations such as IEA or the Energy Charter Treaty. It will be more difficult to influence the behaviour of some key supplier countries where EU leverage is limited. It would be an illusion to believe that the EU can induce them to change their behaviour - often against their interests. Trying to force Russia to sign the Energy Charter Treaty is just such an example.

As a first step, an EU-wide consensus will need to be found, followed by the integration of external energy policy into other policies. Again, the onus will be mainly on the European Commission to ensure coherence and continuity, although Council cooperation in coherence matters has been found important but difficult to obtain (Egenhofer et al., 2006). This will however mean that only those issues on which a consensus exists will be brought to the agenda of summits with third countries. This will probably fall considerably short of assisting oil and gas companies operating in Europe to obtain access to reserves by lending diplomatic weight to EU investors. Such support is likely to be continued by member state governments to companies of national parentage in the absence of a truly integrated EU-wide internal market and a common foreign policy.

5.2 Outlook

There has been institutional progress through the creation of the network of energy correspondents to strengthen the early-warning capacity and coordination to an extent. Further impacts can be expected of the Lisbon Treaty, which retains the article on energy policy introduced by the draft Constitutional Treaty (see Box 1). It can be expected that the 'solidarity clause' will lead to some more concrete measures. In addition, the creation of an EU energy policy competence will offer the possibility to better coordinate action within the Council Secretariat.

If EU member states agreed that 'speaking with one voice' is an objective to pursue, the EU – possibly through the European Commission – should be given some mechanism to assess potential impacts of national *external* energy and security of supply policies for the EU, its internal market and other member states. This could increase transparency of and awareness for the impact of member state policies on the EU and its member states.

The added-value of a High Official for Foreign Energy Policy ('Mr Energy'), as recently proposed by the European Parliament (2007) and some member states, is doubtful. Responsible for coordinating all policies under the scope of the common European foreign policy on energy, such an EU representative would merely duplicate available capacity to represent European interests (or future common positions) to third parties through the Commission or the EU Presidency. Improved coordination can be achieved within the current institutional set-up or at least after the Lisbon Treaty is in place. Improving the current structure would also avoid creating a new bureaucracy and the potential fighting for competence.

6. WHAT PRICE TO PAY FOR ENERGY SECURITY AND SUSTAINABILITY?

In recent decades, energy was abundantly available and at low prices. This situation has changed. Various trends have transformed the global energy market from one favourable to consumers to a sellers' market. With EU domestic reserves rapidly decreasing and tight production capacity, future energy security may require paying a premium. This would also square with analysis by the International Energy Agency (IEA), showing that there are sufficient - but more costly - recoverable resources of fossil fuels, such as deep-water, super-deep and arctic oil, enhanced oil recovery, heavy oil bitumen, oil-shale, and gas or coal to liquids - to cover global demand at or below current oil prices, even if CO_2 costs are included (Pflüger, 2006; IEA, 2006b). At present levels of production and demand, proven reserves far exceed annual demand. [25] In addition, there are renewable, nuclear and clean coal options. IEA analysis also indicates that price increases affect demand; the oil price shocks in the 1970s and 1980s triggered a major demand effect in the form of considerable energy efficiency improvements, the effect stretching over a decade. Similar developments can be observed since the latest oil price rises (Gros et al., 2006). In terms of sustainability, the main challenge will be to deal with considerable costs related to the mitigation of and adaptation to climate change.[26]

---

[25] For instance, the IEA (2006c) estimates that the world reserve base of coal will last for close to 170 years, if the total 2004 proved recoverable reserves are divided by the total production of coal in 2005.

[26] UNFCCC (2007) estimates that an additional $248-381 billion (2005 USD) would be required in the year 2030 to return global GHG emissions to the level of 2004.

Theoretically, achieving both energy policy and climate change objectives could be achieved by true internalisation of external costs. However internalisation of externalities is not a straightforward matter and is fraught with methodological and data issues. In addition, internalisation is not an automatic guarantee for environmental sustainability or security of supply due to market and non-market barriers.

6.1 The cost of security of supply

It is generally assumed that the market price reflects the security premium, assuming that energy supply companies internalise security of supply concerns to a large extent with government regulation doing the rest. Market prices would then reflect all relevant information; at least concerning the public-domain. [27] This view, however, is not shared universally. Market participants tend to give more weight to short-term rather than long-term aspects. Short-term aspects in the oil and gas market are interrelated boom-bust upstream and downstream investment cycles on the one hand and strongly bullish and bearish price expectations on the other. It has therefore been argued that *long-term* energy security risks from a societal point of view may not be adequately factored in prevailing gas and oil prices. The principal reasons are the increasing market power of key oil and gas exporting countries and their possibility to exert windfall rents, political instability and political risks but also uncertainty about the actual rate of depletion of ultimately recoverable oil and gas reserves throughout the world.

Various approaches have been used to assess the external costs of security of supply. Some studies (see Constantini & Gracceva, 2004 for an overview), e.g. by DRI and the US Department of Energy (DoE), have concentrated on GDP losses in the event of a supply crisis, which usually translates into higher oil prices.[28] However, there is no clear correlation

---

[27] It is argued that currently high oil prices reflect a 'risk premium' expressing the political risks associated with a number of key supplier countries.

[28] Constantini & Gracceva (2004) find that the estimates of price increases per barrel due to supply disruptions range from $1.50-$2 for the DRI model to $3.50-$4 for the DoE model, in the event of a sudden cut of oil production of 1mb/d within about three months, assuming oil prices in the $15-20 range. They also find that in 2002, the European Commission estimated that an increase of $10 per barrel of oil is likely to reduce economic growth in industrialised countries by 0.5%. Ogden et

between initial losses and resulting oil price increases because during past supply disruptions, losses were in most cases offset by production increases elsewhere. Other studies, mainly concerned with electricity supply disruptions have attempted to identify the social costs of an electricity supply disruption. Analysis has shown that this value is highly dependent on the quality and composition of various factors, such as population density, duration and continuity, the time of the disruption, the season or the availability and timing of advance warning (Constantini & Gracceva, 2004). ECN have used risk premiums to express security of supply risks (see Jansen & Bakker, 2006). OXERA (2003) has assessed the non-market value of generation technologies such as wind and nuclear, to identify the security of supply premium for different technologies. Finally, a study by the CPB Netherlands Bureau for Economic Policy Analysis (2004) has found that measures to reduce the costs of disturbances on energy markets are generally smaller than costs of measures directed at preventing or mitigating consequences of those disturbances.

6.2 Climate change: More than just a carbon price

Two market failures are generally referred to in the context of climate change.[29] First and foremost, the cost of global warming is not borne directly by GHG emitters, leading to fossil-fuel prices that are 'too cheap' and, as a consequence, to excessive levels of GHG emissions. Second, there are market deficiencies related to the development and adoption of new technologies.[30] Due to *'knowledge spillovers'*, innovating firms cannot keep other firms from benefiting from new knowledge and, therefore, cannot capture all the benefits of innovation for themselves. Also, there are *'adoption spillovers'*, which describe the fact that the cost or value of a new technology to one user may depend on how many other users have adopted the technology. Furthermore, market shortcomings arise due to

---

al. (2004) argue that military expenses to safeguard access to Middle East oil can be used to make a conservative estimate of energy security external cost. They come to $15-$44 per barrel for the US.

[29] This section draws on the summary in de Coninck et al. (2007).

[30] These technology market problems are not as relevant for environmental problems addressed over the course of years as they are for climate policy developing over decades or centuries and requiring much more dramatic changes in technology (see Jaffe et al., 2005, for an overview).

incomplete information. While all investment is characterised by uncertainty, the uncertainty associated with the returns to investment in innovation is often particularly large. Finally, *incomplete information* can be a barrier, for example, when a builder or landlord chooses the level of investment in energy efficiency in a building but the energy bills are paid by a successive owner or a tenant. Hence, the economist's policy prescription to 'put a price' on GHG emissions, thereby forcing individuals and firms to internalise the cost that they are placing on everyone else when they emit GHGs, does not always work.

In all likelihood, the price mechanism will need to be complemented by other measures, such as:

- **Supporting the introduction of new, promising energy-production technologies.** For example, in the case of wind energy, the IEA estimates that each doubling of capacity can lower the costs by 18-20%.[31] Such 'learning curves' differ for different technologies, mainly depending on their maturity. Nevertheless, learning curves have a considerable impact in reducing the cost of GHG mitigation policies in the future. Although present emissions reductions may be relatively modest, certain technologies may contribute significantly to GHG emissions reduction in the future. However, technologies that are far from being competitive should be supported by traditional R&D funding.
- Similarly important is support of long-term **R&D for yet unknown breakthrough technologies.** Markets generally do not provide sufficient incentives for R&D for technologies with a very long time horizon and/or uncertain outcome. Public support will be needed.
- **Investments in energy efficiency** tend to yield benefits in terms of security of supply, climate change and competitiveness, whether they improve insulation or make the choice of efficient appliances or fuels more enticing.[32] Yet, such investments are often not made because of numerous market and non-market barriers, especially in the domestic and small business sectors. Energy efficiency investments have the

[31] See Pflüger (2006).

[32] For example, a recent CEPS/ECN study on a cost-benefit analysis of different climate change options suggests that the level of household expenditure for energy efficiency is lower than justified by net private and social benefits (Egenhofer et al., 2006).

additional value of reducing the total energy bill of households or firms. The latter is particularly important as new energy technologies or policies to combat climate change are expected to increase both wholesale and retail energy prices. More efficient use of energy will be critical as a possible compensation for higher-unit energy costs to keep overall energy expenditure stable.

7. MARKET-BASED POLICY INSTRUMENTS AND TARGETS: HOW, WHY AND WHEN?

A central part of the integrated EU climate and energy policy is to push climate-friendly technologies both to reduce GHG and to foster technological development. At the same time, the EU's commitment to a unilateral approach has increased fears of reduced global competitiveness. The EU emissions trading scheme (ETS) and targets are thus in the focus of the debate, which is also reflected in this Task Force Report.

7.1 The EU ETS

Although the EU ETS has principally been implemented as an instrument to achieve climate change objectives, it can also be seen as a first major attempt at the EU level to internalise an environmental externality (i.e. CO_2 emissions). In theory, under the ETS, the market price of carbon is driven by a combination of the marginal abatement costs of all controlled sources and the emissions cap, thereby ensuring that the environmental objective is achieved at the least cost, thereby minimising the impact on competitiveness. The resulting market price was expected to create long-term predictability, which is critical for spurring investment, while offering flexibility to companies to choose the most cost-effective compliance strategy.

The basic idea of the ETS, i.e. the introduction of a tradable permit system for major fixed installations and the power sector, is generally accepted by most stakeholders in the EU. Stakeholders also agree that the

ETS has a number of shortcomings.[33] These include the lack of investment incentives, possible distortions of competition in the internal market, overall complexity, and its effects on the short-term and long-term competitiveness of industry operating in Europe, mainly as a result of the indirect impact of increased power prices, while its general effectiveness is limited (e.g. Egenhofer, 2007b). If these shortcomings can be fixed, as the revised Commission's EU ETS proposal attempts to do, the EU ETS would potentially be able to internalise the external effects of CO_2 emissions and other GHG emissions, at least to some extent. On the global level, there have been proposals to use the EU ETS to link emissions trading schemes in various countries. So far, however, this has not been pursued, partly because of fears of loss of competitiveness in the absence of a global carbon constraint, and partly because it is difficult to assess the external costs of climate change. In reality, the EU ETS reflects abatement costs. [34]

This leaves open the questions of which instrument is best for dealing with security of supply, and whether market-based or non-market-based instruments are the best options. A few tentative remarks can be made in this regard, as follows:

- To the extent that renewables are seen as contributing to security of energy supply, the motto **'one goal, one instrument'** should be observed.[35] This means that a policy instrument other than the EU ETS should be used to advance the cause of energy security of supply, although instruments should be compatible and ideally mutually reinforcing.

---

[33] See European Commission (2006g), and for an ov.erview, see Egenhofer et al. (2006) and the European Climate Change Programme (ECCP) reports.

[34] The Stern Review (Stern, 2006) on the economics of climate change seems to use estimated marginal abatement costs in the range of \$5 to \$49/tCO_2.

[35] It is a basic principle of economics that sound policy requires at least as many types of policy instruments as there are market problems to be addressed (Tinbergen, 1956). Hence, the optimal set of climate policies would also include instruments explicitly designed to foster innovation, and possibly technology diffusion, in addition to GHG emissions abatement policies that stimulate new technology as a side effect of internalising the GHG externality. Likewise, long-term R&D alone is not sufficient because it provides no direct incentives for the adoption of new technologies and focuses on the longer term, missing near-term opportunities for cost-effective emissions reductions (Philibert, 2003; Sandén & Azar, 2005; Fischer & Newell, 2004).

- **Price signals cannot solve everything**. It is well established that energy efficiency and conservation suffer from a host of market failures mainly in households and the small business sector. These can be addressed by regulation, subsidies, green procurement, labelling or awareness-raising activities. This is especially the case for R&D support for technologies that are far from competitive.
- There may also be a role for environmental taxation as an alternative price mechanism, which may entail lower costs for the covered sectors. Given the unanimity rule, which governs such decisions in the European Council, **taxes are more likely to be applied at member state level**.

7.2 Targets

In discussions over EU energy policy, the concept of targets has been very prominent.[36] This is not surprising. Targets have been an important part of the EU tool box, starting with the EEC Treaty that set a target for creating a Customs Union. A similar approach was used to create the internal market by 1992. As it is generally assumed that this so-called '1992 internal market programme target' has been met, it is considered a success. On various occasions since 1992, the EU has attempted to apply this same formula, e.g. the sectoral targets formulated for renewable energy for electricity and biofuels. In reality, however, the EU record on targets is mixed at best. Witness the difficulties to make progress towards achieving the Lisbon target to transform Europe into "the most competitive and dynamic knowledge-driven economy by 2010", or meeting the Maastricht criteria related to the eurozone or targets agreed for renewable energy for electricity or biofuels in transport.

---

[36] For example Energy Commissioner Andris Piebalgs has raised the question of the wisdom of setting a number of long-term *'sectoral'* targets – i.e. applied to different technologies – such as for energy efficiency or intensity, renewables, the EU ETS but also for technology projects. One could also think of targets for combined heat and power (CHP), carbon capture and storage (CCS), strategic gas storage or energy security as developed by ECN and the Clingendael Institute, which have been developing such targets based on indicators for diversification or demand-side flexibility (Scheepers et al., 2006). Commission President José Manuel Barroso has speculated on the use of an EU target for low-carbon energy within the EU.

Proponents argue that targets can be useful in achieving policy objectives or moving sectors in a certain direction. Even if they are not fully met, the result would still be better than what would have been obtained in their absence. Opponents argue that targets segment the internal market, if the targets are national or differentiated between member states, while encouraging rent-seeking by lobbies (Michaelowa, 2004; Fullerton & Metcalf, 2001). In addition, if targets are too ambitious, there is a risk of imposing excessive costs, making targets politically untenable, possibly leading to a reversal of policy. Under such a scenario, the international credibility of the EU in the area of climate change might suffer as a consequence.

7.2.1 *Best practice in target-setting*

Targets are a good means to express a vision of what the EU and its member states hope to achieve. If properly set, targets can be useful in steering investment in a certain direction while avoiding the most important risks: market segmentation, rent-seeking, lack of credibility or reversal of a policy leading to stranded investment. The following section draws on previous analysis (Egenhofer, 2007c).

Since targets constitute a significant intervention, they should only be applied in cases where the objective to be achieved is significant, e.g. energy security or long-term climate change. The more targets there are, the less flexibility there is for the market to allocate resources. Ultimately, targets can degenerate into 'planning' tools, overriding market incentives and reinforcing rent-seeking. At the same time, objectives must be realistic and achievable, i.e. 'what is needed' should be broadly in line with 'what is possible'.

Targets can only fulfil one objective and not several at a time, although all costs and benefits should be factored in when targets are set.[37]

It is critical to formulate the target properly. By defining the required outcome (e.g. near-zero carbon power generation) rather than prescribing the possible solutions (e.g. renewables, nuclear, CCS, etc.), the market will

---

[37] E.g. boosting renewables to bring down technology costs, increasing biofuels to increase supply flexibility for transport fuels or achieving a certain number of CCS projects for demonstration purposes. For a full analysis of the costs and benefits of different climate change options, see Egenhofer, Jansen, Bakker & Jussila Hammes (2006).

be able to choose the most economic solution. Sectoral targets (e.g. for renewables or CCS) may still be needed for a transition period, to avoid crowding out certain technologies.

*Box 3. A critical analysis of the European biofuels target**

The EU has committed itself to increasing the share of biofuels in transport from the current level of less than 2% to a minimum of 10% by the year 2020. This target is the subject of growing concern, however, as doubts about the environmental and economic efficiency of biofuels are increasing (e.g. House of Commons, 2008).

There is clear evidence that the transport sector should contribute to the EU's climate policy due to its continuing growth related with growing emissions. Biofuels form one part of the EU's strategy to reduce emissions in this sector, but they are subject to growing criticism. However – apart from enhancing vehicle efficiency, which is increasingly expensive the higher the reduction efforts – biofuels are at present the only option for using renewable energy in the transport sector. For stationary applications outside of the transport sector, however, other renewable energy sources are available, such as solar power.

The criticism of biofuels focuses on three main fronts. First, land use changes associated with increasing production of biomass for the use of renewable energy generation may speed up the destruction of natural habitats and lead to large amounts of carbon actually being released from sinks, depending on what land is substituted. Second, using food material for the production of bioenergy could increase food prices with negative implications especially in developing countries. Third, while biofuels have the potential to reduce greenhouse gas emissions, abatement potentials at a lower price are available, especially in the sectors covered by the ETS.

The European Commission proposal for a Directive on renewable energy sources published on 23 January 2008 touches on all of these issues: The greenhouse gas balance and impacts on biodiversity as well as food supply problems have to be resolved by sound sustainability certification. However, the current proposal was criticised for failing to protect important ecosystems and to respond to questions of how to deal with the indirect effects of biofuel production and their impact on developing countries. Given that the EU Council subjected the biofuels target to the condition of their "production being sustainable" (European Council, 2007), several environmental NGOs are asking for the Directive to be improved with "proper safeguards" and for the suspension of the mandatory biofuels target.** German, British or Dutch biofuel sustainability criteria could serve as models for such safeguards.

The questionable cost-benefit ratio of biofuels has to be discussed in a differentiated manner, which is reflected in another precondition mentioned by the EU Council. Large-scale application should be contingent on the development of so-called 'second-generation' biofuels, which can be produced from almost any form of biomass, including agricultural waste and non-food plants. They leave a smaller carbon footprint and perform far better in environmental and economic terms than conventional biofuels. Currently, their development is in its early pilot phase. With respect to its enormous potential in terms of yields per hectare, GHG savings and sustainability, additional funding during further development and deployment will be needed. However, it is unlikely that they will become a viable alternative in the short term. Filling the gap with low-performing first-generation fuels raises both sustainability concerns and problems with fuel quality standards.

Given these uncertainties, a separate target for biofuels in transport seems unjustified. Nevertheless, technologies with high potential for sustainable bioenergy production, such as second-generation biofuels, should be supported and brought to the market.

* See also Behrens (2008).

** Letter of 17 environmental NGOs to the European Commission, dated 9 January 2008.

Targets can be set at the member state or at the EU level. From an internal market and allocation perspective, targets would be ideally set at the EU level, but the current heterogeneity of member states in terms of economic development, the structure of the energy sector - particularly in power generation - and national preferences, makes this more difficult. If targets are expressed at member state level, efficiency and flexibility of implementation can be enhanced by making quotas or obligations tradable across borders, as proposed by the Commission in the form of a renewable energy trading regime.[38] To avoid a race for subsidies, subsidised products or services can be excluded from trading (as proposed by the Commission for renewable energy). Similarly, aligning overall levels of subsidy for the same product or service in each member state would avoid taking 'double advantage' of the system.

If the deadlines for achieving targets are set on an excessively short-term basis, they can create rigidities, whereas long-term targets can lack credibility. Long-term targets are political by definition, in that they express a political ambition on the part of a generation of politicians. The

---

38 COM(2008) 19: "Proposals for a Directive of the European Parliament and of the Council on the promotion of the use of energy from renewable resources".

fact that such targets tend to be ambitious and are seldom based on an economic analysis of the costs and benefits of different options can mean that they may turn out to be excessively costly. This can undermine political acceptability with the result that the next generation of politicians will abandon them. In extreme cases, this can result in 'boom-and-bust' cycles, as observed in the field of renewables in the US. The credibility of targets should therefore increase as they become more realistic, i.e. achievable.[39] The level of credibility will be exposed if progress is tracked, e.g. by the European Commission.

One way of potentially overcoming this credibility gap while allowing for ambitious targets in the EU would be to formulate a set of minimum targets or obligations at EU level, with individual member states being free to go further if they wished.[40]

Ultimately, targets can only make a difference if they are backed up by implementation strategies. Long-term targets provoke trade-offs (e.g. between competitiveness of industry and climate change targets) that need to be settled. These trade-offs are more likely to be addressed when the detailed implementation strategies are formulated by the EU or by member states.

---

[39] An important additional factor accounting for the credibility of a target is whether it makes allowances for member state differences and preferences, e.g. resource endowment or CO_2 intensity.

[40] The UK and Sweden, for example, have done precisely this in the area of climate change targets.

8. Case Study: Long-Term Targets within an EU Framework for Renewable Electricity Support

Support for renewable energy is one of the most delicate issues within the integrated EU climate and energy policy. Setting long-term targets (until 2020) is as important as the harmonisation of support schemes. Long-term targets are important to give investment signals for specific periods and thus enhance regulatory stability and consistency, in addition to providing transparency to investors. Harmonisation of (divergent) national support schemes among EU member states is warranted as a result of increasing levels of renewables. A greater percentage of power produced by renewables (RES-E) – coupled with different member state stimulation policies – is likely to result in distortions to Europe's renewable electricity markets, essentially raising barriers to trade and affecting competition. Policy harmonisation is also expected to increase the stability of the electricity grid and to allow for economies of scale through the creation of liquid and efficient markets.

8.1 Harmonisation and coordination of renewable support

In a transition period, competition between national support schemes can be healthy. Member states exploring different options may lead to a greater variety of solutions from which to choose. As long as volumes remain relatively low, cross-border externalities (e.g. negative impacts on other member states or the EU) remain limited, unless national schemes erect barriers to trade or distort competition. As volumes of renewables increase, the cross-border effects on for example competition in the power sector or grids will increase. In addition, there may be efficiency gains through scale effects. Over time, this will most likely increase the rationale for EU-wide convergence of support schemes, if not outright harmonisation. Based on previous analysis (e.g. Jansen et al., 2005; Egenhofer & Jansen, 2006), we separate the entire sphere of national support-scheme frameworks into three distinct parts: i) level of support, ii) support-scheme models and iii) the legal framework including regulatory issues, and identify possible steps towards full harmonisation. A summary is provided in Table 2.

Level 1: Coordination of support levels

This basic level of coordination aims to avoid the most important distortions in member states' signals for renewable projects. The level of support - direct and indirect - has a direct impact on decisions on whether to invest on one or the other side of a border. As EU-wide harmonisation of support systems does not seem a realistic prospect in the short or medium term, policy-makers should have a close look at the impacts of different levels of overall support across member states and their effects at the location around borders. It should be expected that reliable information on total support levels in terms of €/MWh per renewable generation technology will reveal whether support schemes seriously distort investment decisions in border areas.

Level 2: Harmonisation of support schemes

Another, more difficult area is harmonisation or coordination of the different models of support schemes. Member states use a variety of different support schemes, which are often based on national preferences and policy choices. At the moment there is no consensus within the EU as to which of the existing models should become the blueprint of an EU-wide scheme, e.g. feed-in tariffs, green certificate systems, tendering systems or tax incentives. In addition, many renewable projects still enjoy guarantees under existing support schemes.

Until such a consensus is feasible, stocktaking and information dissemination of best-practice design features can be an important step. Further, it is imperative that reliable, standardised data are collected on aggregate support levels to promote the market uptake of distinct technologies in each member state. This enables policy-makers to compare the intensity of policy efforts on the one hand and to expose undue market distortions on the other, allowing for differences in national renewable resource endowments.

In this context, a major question is whether and if so, when to link schemes that are based on portfolio standards and allow for trading of certificates. Economically speaking, making certificates tradable will increase efficiency as the target will be achieved at the least cost. Two caveats need to be made, however. First, some member states are still in the process of experimenting with their support systems. As a result, they may still be struggling with design issues. Linking efficient support schemes to inefficient ones may undermine the overall scheme. Second, it is important to consider distributional impacts of costs and benefits. One of the rationales behind the support of renewables has been that it would permit the EU to gain a competitive edge in a 'sunrise' technology. Politicians have used this argument in the past to justify support either via subsidies or by consumers under portfolio schemes. Under a cross-border or EU-wide portfolio model, renewable sources will be developed in the most suitable location (e.g. wind in Scotland). This ensures least-cost compliance with the target. On the other hand, this may erode support by those member states that due to resource endowment do not benefit from renewables investment. As a result, such member states will most likely only accept a modest target. Thus, while the overall target may be reached at least-cost, the overall (aggregate) EU target may be lower than it otherwise would be. This is a trade-off.

Nevertheless, member states should be free to link their schemes, if they wish. A precondition for linking would be the introduction of renewable energy guarantees of origin (RE-GO). The preparation of harmonised RE-GO schemes should receive priority.

Level 3: EU-wide regulatory framework for support

The third level of harmonisation of support schemes is the creation of an EU-wide regulatory framework for support. While many aspects will remain within the responsibility of the member states (e.g. administration, permitting), the implementation of renewables support policy will need to be undertaken within a common EU framework. Different elements of this framework can be developed within different timeframes. The following tentative timeframe for coordinated action at the EU level is recommended for further consideration (Table 2).

Table 2. Timeframe for action

Action	Timing of introduction		
	By 2008	Before 2010	After 2015
Analysis of best practises of support schemes*	√		
Harmonisation of support mechanisms			
a) Total level of support in border areas	√		
b) Support schemes		√ **	
c) EU framework		√	
- Removing mandatory support			√
- Preparation of harmonised RE-GO schemes	√		
- Introduction of harmonised RE-GO/GO schemes		√	
Grid integration			
a) Grid extension planning	√		
b) Coordination and information-sharing among regulators and TSOs	√		
c) Harmonisation of member state grid access codes and standards for network equipment		√	
d) RES-E priority dispatch re-examination		√	
Authorisation procedures			
a) Promotion of streamlining of authorisation procedures based on best practice	√		
b) Promotion of regional one-stop authorisation	√		

* This action assumes agreed data collection.

** Given a 7-year transition period, the implementation of harmonisation decisions will fully come into operation by 2015 at the earliest and 2017 at the latest.

Source: Jansen, Gialoglou & Egenhofer (2005); Egenhofer & Jansen (2006).

Enhancing security of supply further

From the perspective of both enhancing the security of energy supply and promoting sustainable development, the EU has embarked on constructive engagement with neighbouring countries. Neighbouring countries that are willing to fully transpose the Renewables Directive onto their respective national legislation, including the adoption of indicative targets to be agreed upon with the European Commission, should be allowed to share not only its obligations but also its benefits. For example, neighbouring countries with large endowments of RES-E (hydro, wind and sun) should countries with large endowments of RES-E (hydro, wind and sun) should be allowed to transfer RE-GOs to EU member states for disclosure applications. This action could provide additional revenue streams for, and facilitate financing of, renewable-electricity generating projects in these countries. This policy may apply to EEA countries (notably Norway) and to Switzerland (through its bilateral agreements with the EU), yet equally well in the medium term to countries that are part of the European Neighbourhood Policy area.

References

Baumol, W.J. and W.E. Oates (1988), *The theory of environmental policy* (2nd edition), Cambridge: Cambridge University Press.

Behrens, A. (2008), "Biofuels or bicycles? Why the European Union should reconsider its biofuels target", CEPS Commentary, CEPS, Brussels.

BP (2007), *BP Statistical Review of World Energy*, BP, London.

Constantini, V. and F. Gracceva (2004), *Social Costs of Energy Disruptions*, INDES Working Paper No. 6, CEPS, Brussels.

Council of the EU (2005), *Information Note 7242/05*, 11 March.

CPB Netherlands Bureau for Economic Policy Analysis (2004), *Energy Policies and Risks on Energy Markets: A cost-benefit analysis*, CPB Special Publication 51, The Hague.

De Coninck, H., C. Fischer, R.G. Newell and T. Ueno (2007), *International Technology-Oriented Agreements to Address Climate Change*, RFF Discussion Paper 06-50, Resources for the Future, Washington, D.C.

Egenhofer, C. (2007a), "Integrating Security of Supply, Market Liberalisation and Climate Change", in M. Emerson, *Readings in European Security*, Vol. 4, Centre for European Policy Studies, Brussels, pp. 79-96.

Egenhofer, C. (2007b), "The Making of the EU Emissions Trading Scheme: Status, Prospects and Implications for Business", *European Management Journal*, Vol. 25, No. 6, December, pp. 453-463.

Egenhofer, C. (2007c), *Looking for the cure-all? Targets and the EU's New Energy Strategy*, CEPS Policy Brief No. 118, CEPS, Brussels, January.

Egenhofer, C. and K. Gialoglou (2004), *Rethinking the EU Regulatory Strategy for the Internal Energy Market*, CEPS Task Force Report, CEPS, Brussels, December.

Egenhofer, C. and L. van Schaik (2005), *Towards a Global Climate Regime: Priority Areas for a Coherent EU Strategy*, CEPS Task Force Report, CEPS, Brussels.

Egenhofer, C. and J. Jansen (2006), "A timetable for harmonisation of support schemes for renewable electricity in the EU", *European Review of Energy Markets*, Vol. 1 (2), pp. 57-83.

Egenhofer, C., J.C. Jansen, S.J.A. Bakker and J. Jussila Hammes (2006), *Revisiting EU Policy Options for Tackling Climate Change: A Social Cost-benefit Analysis of GHG Emissions reduction* Strategies, CEPS/ECN.

Egenhofer, C., L. van Schaik, M. Kaeding, A. Hudson and J. Núñes Ferrer (2006), *Policy Coherence for Development in the EU Council: Strategies for the way forward*, CEPS, Brussels.

European Commission (2000), *Towards a European Strategy for the Security of Energy Supply*, Green Paper, COM(2000)769 final (http://www.europa.eu.int/comm/energy_transport/en/lpi_lv_enl.html).

European Commission (2001), *European Climate Change Programme*, Final Report (http://www.europa.eu.int/comm/environment/climat/eccpreport.htm)

-------- (2005a), *Winning the Battle against Global Climate Change*, Communication from the Commission to the Council, the European Parliament, the European Economic and Social Committee and the Committee of the Regions, COM(2005)35 final.

-------- (2005b), *Winning the Battle Against Global Climate Change – Background Paper*, Commission Staff Working Paper, Brussels, 9 February.

-------- (2005c), *Doing more with less*, Green Paper on Energy Efficiency, COM(2005)265, 22 June.

-------- (2006a), *A European Strategy for Sustainable, Competitive and Secure Energy*, Green Paper, COM(2006)105, 8 March.

-------- (2006b), *Annex to the Green Paper "A European Strategy for Sustainable, Competitive and Secure Energy – What is at stake?"*, Background Document, Commission Staff Working Document, COM(2006)105, 8 March.

-------- (2006c), *Action Plan for Energy Efficiency: Realising the Potential*, Communication form the European Commission , COM(2006)545, 19 October.

-------- (2006d), *First report of the High Level Group on Competitiveness, Energy and the Environment*, 2 June.

-------- (2006e), *Second report of the High Level Group on Competitiveness, Energy and the Environment*, 30 October.

-------- (2006f), *External energy relations – from principles to* action, Communication form the European Commission to the European Council, COM(2006)590, 12 October.

-------- (2006g), *Building a global carbon market*, Communication form the European Commission, COM(2006)676, 13 November.

European Commission and the Secretary General/High Representative (SG/HR) (2006), *An external policy to serve Europe's energy interests*, S160/06.

European Council (2005), Presidency Conclusions, Brussels, 23 March, 7619/05 CONCL 1.

European Council (2007), Presidency Conclusions, Brussels, 2 May, 72241/1/07 REV 1 CONCL 1.

European Parliament, (2007), *Report on towards a common European foreign policy on energy*, Committee on Foreign Affairs, Brussels.

Fischer, C. and R. Newell (2004), *Environmental and Technology Policies for Climate Change*, Discussion Paper 04-05-REV, Resources for the Future, Washington, D.C.

Fullerton, D. and G. Metcalf (2001), "Environmental controls, scarcity rents, and pre-existing distortions", *Journal of Public Economics*, 80, pp. 249-267.

Gros, D., T. Mayer and A. Ubide (2006), *A World out of balance?*, Special Report of the CEPS Macroeconomic Policy Group, CEPS, Brussels.

House of Commons (2008), *Are biofuels sustainable?*, Environmental Audit Committee, UK House of Commons, London.

IEA (International Energy Agency) (1995), *Natural Gas Security Study*, IEA/OECD, Paris.

IEA (2005), *Resources to Reserves – Oil & Gas Technologies for the Energy Markets for the Future*, International Energy Agency, Paris.

IEA (2006a), *World Energy Outlook 2006*, International Energy Agency, Paris.

IEA (2006b), *Energy Technology Perspectives 2006: Scenarios & Strategies to 2050*, International Energy Agency, Paris.

IEA (2006c), *Coal Information 2006*, International Energy Agency, Paris.

IEA (2007), *World Energy Outlook 2007*, International Energy Agency, Paris.

IISD (International Institute for Sustainable Development) (2007), *Government support for ethanol and biodiesel in the EU*, Global Subsidies Initiative, Geneva.

Jaffe, A.B., R.G. Newell and R.N. Stavins (2005), "A Tale of Two Market Failures - Technology and Environmental Policy", *Ecological Economics* 54, pp. 164-174.

Jansen, J. and S. Bakker (2006), *Social cost-benefit analysis of climate change mitigation options in a European context*, ECN Report No. ECN-E--06-059, ECN - Energy Research Centre for the Netherland.

Jansen, J., K. Gialoglou and C. Egenhofer (2005), *Market Stimulation of Renewable Electricity in the EU: What degree of harmonisation of support mechanisms is required?*, CEPS Task Force Report , CEPS, Brussels.

Luciani, G. (2004), *Security of Supply for Natural Gas Markets: What is it and what is it not?*, INDES Working Paper No. 2, CEPS, Brussels, March.

Michaelowa, A. (2004), *The German wind energy lobby: How to successfully lobby for expensive technological change*, HWWA Discussion Paper 296, Institut für Wirtschaftsforschung, Hamburg.

Müller, F. (2007), *Energy Security – Demands Imposed on German and European Foreign Policy by a Changed Configuration in the World Energy Market*, SWP Research Paper No. 2, SWP, Berlin.

Odgen, J.M., R.H. Williams and E.D. Larson (2004), "Societal lifecycle costs of cars with alternative fuels/engines", *Energy Policy*, 32, pp. 7-27.

OECD (2007), *Biofuels: Is the cure worse than the disease?*, Roundtable on Sustainable Development, OECD, Paris.

OXERA (2003), *The Non-market Value of Generation Technologies*, June.

Pflüger, A. (2006), "The potential impact of new energy technologies and their diffusion", presentation at the second meeting of the CEPS Task Force, Brussels, 8 November.

Philibert, C. (2003), *Technology Innovation, Development and Diffusion*, OECD and IEA Information Paper, COM/ENV/EPOC/IEA/SLT(2003)4, International Energy Agency, Paris.

Renda, A. (2006), *Impact Assessment in the EU – The State of the Art and the Art of the State*, CEPS, Brussels.

Riley, A. (2006), *The Coming of the Russian Gas Deficit: Consequences and solutions*, CEPS Policy Brief No. 116, CEPS, Brussels.

Röller, L., J. Delgado and H. Friederiszick (2007), *Energy: Choices for Europe*, Bruegel Blueprint Series, Bruegel, Brussels.

Sandén, B. and C. Azar (2005), "Near-Term Technology Policies for Long-Term Climate Targets - Economy-Wide versus Technology Specific Approaches", *Energy Policy* 33, pp. 1557-1576.

Scheepers, M., A. Seebregts, J. de Jong and H. Maters (2006), *EU Standards for Energy security of Supply*, ECN/Clingendael International Energy Programme, ECN-C—06-039/CIEP, June.

Stern, J. (2002), "Security of natural gas supplies", *Journal of the Royal Institute of International Affairs*, London, July.

Stern, J. (2006), *The new security environment for European gas: Worsening geopolitics and increasing global competition*, Oxford Institute for Energy Studies, NG 15, October.

Stern, N. (2006), *The Stern Review on the Economics of Climate Change*, Cambridge: Cambridge University Press.

Stouge, A. (2006), "Improving the innovation chain in Europe", presentation at the second meeting of the CEPS Task Force, Brussels, 8 November.

Tinbergen, J. (1956), *Economic Policy: Principles and Design*, Amsterdam: North-Holland.

UNFCCC (2007), *Report on the analysis of existing and potential investment and financial flows relevant to the development of an effective and appropriate international response to climate change*, Dialogue working paper 8, Dialogue on long-term cooperative action to address climate change by enhancing implementation of the Convention, Fourth workshop, Vienna.

ANNEX 1.

Table 1. Classification of security-of-supply risks in the EU by sector – Oil, gas, coal, nuclear, renewable energy sources (RES) and electricity

	Classification	Event	Disruption	Price rise		Probability in 20 years	Duration	Fuel affected					
				Intl.	Domestic			Oil	Gas	Coal	Nuclear	RES	Elec.
Political risks													
1	Export embargo	Embargo of a specific exporter (e.g. Iraq)	Little	Little	Little	High	Months, years	x	x	–	–	–	–
2	Output reduction	Quotas on production to raise prices (e.g. OPEC cartel)	Yes	Yes	Yes	High	Months, years	x	–	–	–	–	x
3	Local market disruption I	By pressure groups (e.g. fuel price protest)	Yes	–	Yes	Medium-high	Weeks, months	x	–	–	–	–	x
4	Local market disruption II	Regulatory shortcomings (e.g. California power crisis, Nordic market)	Yes	No	Yes	Medium-high	Weeks, months	–	–	–	–	–	x
5	International market disruption	Regulatory failure (e.g. regulation, competition and financial markets)	Yes	Yes (or rationing)	Yes	Medium	Weeks, months, years	x	x	–	–	–	x

Table 1, cont.

6	*Force majeure*	Civil unrest, war, deliberate blockage of trade routes	Yes	Yes	Yes	Low-medium	Variable	x	x	–	–	–	–
7	Import embargo	Embargo of importing state by ex- port or transit country (e.g. gas cut-off)	Yes	No	Yes	Very low for EU	Months, years	x	x	–	–	–	–
Economic risks													
8	Public opinion on large-scale investment	Delay in planning, under-investment	Yes	No	Yes	High	Years	x	x	x	x	x	x
9	Supply discontinuity	Lack of infrastructure	Yes	Yes	Yes	Low-medium	Months, years	x	x	–	–	–	x
10	Production discontinuity	Shortage of production capacity	Yes	Yes	Yes	Low	Years	x	x	–	–	–	x
Environmental risks*													
11a	Accidents	– Major oil spill (land or sea)	No	Yes	Yes	Medium	Weeks, months	x	–	–	–	–	–
11b		– Major nuclear accident	Yes	No	Yes	Low	Months, years	–	–	–	x	–	x
11c		– Burst of major gas pipeline	Yes	Yes	Yes	Low	Weeks, months	x	x	–	–	–	x

Table 1, cont.

12a	Disruption/ destruction of habitat	- Massive biomass plantations; - Ultrasonic waves (of wind turbines);	Yes	No	Yes	High	Months, years	–	–	–	–	–	–
12b	Run-away greenhouse effect	Clear indicators in biosphere (e.g. the melting of permafrost)	Yes	No	–	Very low	Perm./ irreversible	x	x	x	–	x	x
Technical risks													
13	System failure	Technical failure, e.g. due to extreme weather conditions, technical neglect	No	No	Yes	Medium	Days, weeks	–	–	–	–	–	x

* Environmental risks are risks to supply only in an indirect way. Risks from accidents or other environmental dangers are related to subsequent government action, which might act as a dampener to investment and therefore create bottlenecks. Strictly speaking, environmental risks could also be listed under political risks.

Source: Adapted from Egenhofer & Legge (2001).

ANNEX 2. MEMBERS OF THE CEPS TASK FORCE AND INVITED GUESTS AND SPEAKERS

Chairman: Knud Pedersen
Vice President
DONG Energy A/S

Rapporteurs: Arno Behrens
Research Fellow
CEPS

Christian Egenhofer
Senior Fellow
CEPS

Monica Alessi
Research Fellow
CEPS

Chris Anastasi
Senior Environmental Advisor
British Energy plc

Birgitte Andersen
Officer in charge of Energy
EFTA European Free Trade Association

Olafia Dogg Asgeirsdottir
Goods Division
EFTA European Free Trade Association

Joachim Balke
Advisor EU Affairs
E.ON AG

Georg Bäuml
Environmental Strategy
Volkswagen AG

Markus Becker
European Energy Policy Executive
GE Energy

Nicole Beha
Assistant
Bundesverband der Deutschen Industrie (BDI)

Gunnar Bengtsson
Director
Environmental Affairs
Volvo AB

Lisa Boch-Andersen
Senior Advisor
European Union Affairs & Media Relations
ExxonMobil Petroleum & Chemical

Gerard Bos
Holcim Group Support (Brussels) SA

Frank Brannvoll
Managing Director
Brussels Office
Point Carbon Brussels

Georg Brodach
Senior Vice President
Asea Brown Boveri Europe Ltd

Reinhold Buttgereit
Director
Vattenfall European Affairs

Massimiliano Calamea
European Affairs Manager
Public & Regulatory Affairs
ENEL SpA

Oscar Cano Arias
Account Manager
Hill & Knowlton International S.A.

Michel Catinat
Head of Unit
DG Enterprise
European Commission

Christian Chavane
Vice President
European Affairs
Total

Jennifer Cosco
Analyst
Goldman Sachs International

Michel Cruciani
Public Affairs
Electricité de France (EDF)

Frits de Groot
Senior Advisor
Environmental Affairs
VNO-NCW (Confederation of
Netherlands Industry & Employers)

Olivier Debande
Senior Advisor
European Investment Bank (EIB)

Veronique Deman
DG Enterprise
European Commission

Bernd Dittmann
Director
Bundesverband der Deutschen
Industrie (BDI)

George Dura
Research Assistant
CEPS

Christine Faure-Fedigan
Director CO_2 Strategy
Gaz de France

Antje Fiehn
Advisor
Energy, Transport,
Telecommunications
Bundesverband der Deutschen
Industrie (BDI)

Pia Maria Funari
Senior Manager
Brussels Office
ENEL SpA

Kyriakos Gialoglou
Desk Officer - responsible for Energy
DG Health and Consumer Protection
European Commission

Per-Olof Granström
Executive Vice President
Svensk Energi - Swedenergy AB

Wim W.P. Groenendijk
Manager
Regulation & Projects
N.V. Nederlandse Gasunie

Ursula Hartenberger
EU Public Affairs Manager
RICS Communications
Royal Institution of Chartered Surveyors 'RICS'

Emmanuel Haton
Assistant Director
Government Affairs
BP Europe

Stephan Herbst
Manager
Environment Strategy
Toyota Motor Europe

Kerstin Hötzel
Education Development Executive
Royal Institution of Chartered Surveyors 'RICS'

Morten Husby
Natural Gas Division
Statoil

Lex Huurdeman
Manager
Regulatory Affairs & Commercial Strategy
Shell International E&P

Esa Hyvärinen
Vice President
Public Affairs
Fortum Corporation

Karl Isaksson
Consultant
KREAB AB

Kuniaki Ito
Resident Executive Director
Europe & Middle East
Japan Bank for International Cooperation

Richard Ivens
Senior Manager
Foratom

Staffan Jerneck
Director & Director of Corporate Relations
CEPS

Hans Korteweg
Head of Cabinet & Deputy Head of Unit
Institutional Affairs
Foratom

Helene Lavray
Government Relations Research Officer
Rio Tinto

Maria José Lopez Quilez
Regulation Analyst
Endesa

Bengt Magnusson
European Correspondent
ERA Magazine
Svensk Energi - Swedenergy AB

Roland-Jan Meijer
Vice-President
Environmental Affairs
Holcim Group Support (Brussels) SA

Wolfgang Mülkens
Abt Energiepolitik
Bundesverband der Deutschen Industrie (BDI)

Lars Nielsen
Counsellor (Energy)
Permanent Representation of Denmark to the EU

Philippe Paelinck
Director CO_2
ALSTOM Power

Bernadett Parrag
Policy Officer
City Office in Brussels

Jean-Paul Peers
Vice President - Energy Policy
Siemens AG

Jennifer Pierotti
Researcher
Japan Bank for International Cooperation

Hugh Porteous
Vice President Corporate Affairs
Rio Tinto (Alcan)

Jouni Punnonen
Senior Advisor
Confederation of Finnish Industries

Lisa Rabbe
Managing Director
Government Affairs
Goldman Sachs International

Adhir Ramdarshan
Junior Consultant
Public Affairs
Fleishman-Hillard Company

Birgitta Resvik
Director
Energy & Climate Change Policies
Confederation of Swedish Enterprise

Olivier Ricard
European Financing Delegate
European Affairs
Total

Armin-D. Riess
Deputy Head
Economic & Financial Studies
European Investment Bank - EIB

Paul Rübig
MEP
European Parliament

John Scowcroft
Head of Unit
Environment & Sustainability
Eurelectric

Emmanuel Sellier
Head of EDF Liaison Office
Mission Europe
Electricité de France (EDF)

Stephan Singer
Head of Unit
Climate Policy
WWF European Policy Office

Anna Sole-Mena
Administrator
DG Entreprise
European Commission

Kaisa-Maria Soro
Associate Consultant
KREAB AB

Thomas Spiller
Director, Public Policy EMEA
SAS

Alexey Spirin
Project Director
Kyoto Protocol
RUSAL

Damjan Stanek
Director
Borzen

Rainer Steffens
Minister Counselor
Environmental Policy
Permanent Representation of
Germany to the EU

Francesca Stevens
Confindustria

Lars B. Stoltenberg
Energy Policy Manager
Hydro Oil & Energy

Jan Sundell
Senior Adviser
Vattenfall

Alastair Sutton
Partner
White & Case LLP

Eva Svitakova
EU Specialist
European Affairs
CEZ Group

Marie Thiellesen Bjerborg
Danish Energy

Gabor Toth
Adviser
Unit Energy Policy & Generation
Eurelectric

Hervé Touati
Senior Expert
McKinsey & Company

Jonny Trapp Steffensen
Advisor
DONG Energy A/S

Sami Tulonen
Director
Institutional Affairs
Foratom

Matti Vainio
ECHA - Helsinki

Juan Eduardo Vasquez Moya
Deputy Manager
International Regulation
Endesa

Stefaan Vergote
DG Environment
Climate Change Unit
European Commission

Jean-Arnold Vinois
Head of Unit
DG Transport & Energy
European Commission

Peter Vis
Member of Cabinet
Commissioner Andris Piebalgs
European Commission

Ruud Wassen
Public Affairs
Fleishman-Hillard Company

Yusuke Watanabe
Senior Representative
Japan Bank for International Cooperation

Wolfgang Weber
Head of Energy Policy
Energy Unit
BASF AG

Ramus Wendt
Rapporteur on the EU-South Korea FTA, Business Europe
Confederation of Danish Industries (DI)

Manfred Wiegand
Partner
Global Utilities Leader
PriceWaterhouseCoopers

Klaus Willnow
Director Public Affairs Energy
Siemens AG

Markus Wolf
TTC-TT Technology Centre
ALSTOM Power

Sophie Wozniak
Consultant
Hill & Knowlton International S.A.

Mike Wriglesworth
Senior Advisor
CEPS

Roberto Zangrandi
Head of European Affairs
ENEL SpA

Invited Guests and Speakers

Riku Huttunen
Deputy Director General
Ministry of Trade & Industry,
Finland

Peter Kernan
Managing Director
Energy Analytical Group
Standard & Poor's

François Nguyen
Senior Policy Advisor
International Energy Agency (IEA)

Antonio Pflüger
Head
Energy Technology Collaboration
Division
International Energy Agency (IEA)

Asta Sihvonen-Punkka
Chair Electricity WG,
Council of European Energy
Regulators (CEER)

Anders Stouge
Director
Confederation of Danish Industries
(DI)

Jean-Arnold Vinois
Head of Unit
DG Transport & Energy
European Commission

Luc Werring
Head of Unit
DG Energy & Transport
European Commission